Einstein based his theory on two simple postulates:
1 The principle of Relativity
2 The Constancy of Speed of light in vacuum

From the infinitely small to the infinitely big covering over
sixty spatial orders of magnitude quantum theory is used as
much to describe the still largely mysterious vibrations of the
microscopic strings that could be the basic constituents to
explain the fluctuations of the radiation reaching us from the
depths of outer space.
Serge Haroche

Table of Contents

Chapter 1: The Copenhagen Interpretation   6

- The Deflection of light   8

- Michelson-Morley Experiment   11

- The displacement of the Position   14

- Rotational sources   16

- Low Density Inflationary Universes   17

Chapter 2: Light in Contact with Matter   24

Pair Production   26

The Stern-Gerlach Experiment   31

CQED   42

Chapter 3: Quantum Mechanics   68

- Quantum Mechanics   69

Chapter 4: The Density Probability

The Copenhagen Interpretation

The Copenhagen Interpretation

The essential controversial features of the Copenhagen Interpretation are:

1 The Uncertainty Principle of Heisenberg

2 The Principle of Complementarity of Bohr

With the passage of time the Copenhagen Interpretation has been identified with a concept known as the collapse of the wave function so called the reduction of the wave packet as formulated by Von Neumann.

The Heisenberg Principle

Heisenberg Uncertainty principle asserts that the product of position and momentum for any particle will be more than a certain multiple of Planck's constant. Since momentum is the product of mass and velocity, and since the mass in experiment is usually that of an electron, this can equivalently be described by saying that the more precise the position of a electron is known, the less precise will be known of its velocity.

The Bohr Principle

Bohr Principle of Complementarity arose out of the difficulty physicists were having in their attempts to determine whether quantum phenomena such as light are particles or waves. But complementarity is not a solution. Instead, it is an assertion that no solution exists and that physicists know all that can be known about the question. So it would be well to examine the context from which complementarity arose.

The Deflection of light

Theories of the deflection of light by mass date back at least to the late 18th century. At that time, the Reverend John Michell, an English clergyman and natural philosopher, reasoned that were the Sun sufficiently massive, light could not escape from its surface. The pioneer of a mathematical description of gravity, Sir Isaac Newton, apparently wrote nothing about the effect of mass on the path of light rays, other than to note at the end of his treatise, "Opticks," published in 1704, that light particles should be affected by gravity in the same way as is ordinary matter. The first calculation of the deflection of light by mass was published by the German astronomer Johann Georg von Soldner in 1801. Soldner showed that rays from a distant star skimming the Sun's surface would be deflected through an angle of about 0.9 seconds of arc, or one quarter of a thousandth of a grad. This angle corresponds to the apparent diameter of a compact disc viewed from a distance of about 30 kilometers nearly 20 miles. Soldner's calculations were based on Newton's laws of motion and gravitation, and the assumption that light behaves like very fast moving particles. As far as we know, neither Soldner nor later astronomers attempted to verify this prediction, and for good reason: Such an attempt would have been far beyond the capability of early 19th century astronomical instruments. Over a century later, in the early 20th century, Einstein developed his theory of general relativity.

Deflection of light passing close to a massive body

According to general relativity, a light ray arriving from the side would be bent inwards such that its apparent direction of origin, when viewed from the other side, would differ by an angle α; the deflection angle whose size is inversely proportional to the distance d of the closest approach of the ray path to the center of mass. The curves to each side show the dependence of the deflection angle α on the distance d. The deflection angle is largest when the light rays pass closest to the mass; α becomes smaller as d becomes larger. For the Sun, the curves look similar, but the predicted value of α is five thousand times smaller for rays that skim the surface of the Sun than for rays that skim the surface of this pseudo Sun.

Shifting positions

An observer on the Earth can detect the deflection by the Sun of the light from a distant star by the change with time of year in the star's apparent position in the sky. In the absence of a mass, the light follows a straight line from the star to the observer. In the presence of the mass, the light ray is bent, and the light reaches the observer from a slightly different direction. This direction defines a star's apparent position in the sky. Such shifts in position - although far smaller than imagined - should be visible to a properly equipped optical observer on the Earth for a star near the Sun's path in the sky. However, under ordinary conditions sunlight causes the atmosphere to be so bright that it is not feasible to observe from the Earth with optical telescopes any star whose light passes near the Sun. For the first tests of Einstein's predictions, astronomers therefore used solar eclipses, occasions on which the moon is between the Earth and the Sun, and blocks all sunlight from reaching the vicinity of the Earth and brightening its atmosphere.

Measuring light deflection

1919 saw the first successful attempt to measure the gravitational deflection of light. Two British expeditions were organized and sponsored by the Royal Astronomical Society and the Royal Society. Each of the two groups took photographs of a region of the sky centered on the Sun during the May 1919 total solar eclipse and compared the positions of the photographed stars with those of the same stars photographed from the same locations in July 1919 when the Sun was far from that region of the sky. The results showed that light was deflected, and also that this deflection was consistent with general relativity but not with "Newtonian" physics. The subsequent publicity catapulted Einstein to world fame, and led to his having the only ticker-tape parade ever held for a scientist on Broadway New York With repetitions of eclipse measurements over the next half century, astronomers were able to improve on the accuracy of these first results by only about a factor two, yielding a confirmation of general relativity to within about ten percent. The breakthrough came in 1967 with the realization that simultaneous measurements with a set of radio telescopes, "Very Long Baseline Interferometry" could be used to measure light deflection with much greater accuracy. In addition to providing the means to test general relativity to high accuracy, the fact that mass deflects light has been a great boon to studies of the universe. Masses acting as gravitational lenses have now become a standard tool of astronomy. They allow astronomers to infer the masses of cosmic objects, and the structure and size scale of the universe with some caveats. Through their magnifying effect, gravitational lenses have also been used to observe the properties of very distant galaxies and quasars, as well as to search for planets around distant.

Irwin Shapiro is the Timken University Professor at Harvard University and a Senior Scientist of the Smithsonian Institution.
He works at the Harvard-Smithsonian Center for Astrophysics in Cambridge, MA. His research interests involve applications of radio and radar techniques to problems in geophysics, planetary physics, and astrophysics. He also devised and carried out precision tests of general relativity within our solar system.

Steven Shapiro Associate Professor of Physics and Associate Academic Dean at Guilford College in Greensboro, NC. His research interests are in the dynamics of the Earth's upper mantle, seismology, and most recently, precision measurements of light deflection near the sun.

Michelson-Morley experiment
The experiment used a pencil beam of light that was split in

two parts. One travelled directly ahead through a series of mirrors to increase the length of the travel back to an interferometer. The second was directed at right angles to the first, through an identical system of mirrors out and then back to the interferometer. Any sufficient difference, in time taken for one part of the beam as against the other one would show up as an interference. It was argued that sufficient time difference should have resulted to produce interference. The value obtained by substituting the speed of light and the earth orbital velocity was within the sensitivity pattern of the interferometer. In the event no difference was detected however the apparatus was oriented with respect to the direction of travel of the earth. The Michelson experiment had given negative results because travel at high speed induced a contraction in length of apparently solid bodies and that the contraction resulting from travel at Earth orbital velocity, though very small indeed, was exactly the amount needed to contract the apparatus used so as to cancel out the predicted result.

Two conclusions were drawn:

1 Since ether was not detected, the medium ether had not existed

2 As to measurements generated no measurable gap with the Earth direction of movement, light should have been travelled at a constant speed irrespective of the position of the source or observer.

Relative motion

Relative motion is the relationship between two different objects in different inertial states. For instance, a light source moving at a uniform rate of speed relative to a stationary observer, or two such light sources in motion relative to each other, etc. Less understood are the relationships of the motions involving the light source and observer, or two light sources, and the light radiating from such sources. Take, for instance, the light radiating at speed c in all directions from a light source that is moving at speed v relative to an observer. In considering such examples it is rather obvious that such an observer can only speculate on the light radiating in all directions from such source based on the light that radiates in the direction of that observer. More specific observer can only detect the light that radiates directly to that observer's present point in space. Then upon applying the single point of existence principle to each of the three entities, i.e. source, light, and observer, it is apparent that the observer and detected light are at the same location in space at that instant relative to the source which is at some other location. Obviously, then, the source is not at the location it was when the detected light was emitted, but is at some other location at the exact instant the light is detected. Thus, these two points, the source and observer locations are each at a single location in all the frames of reference at the instant of detection. The point of emission, however, varies from frame to frame depending on the relative motion between frames. We can show the source and observer in the same spatial relationship to each other in all the frames of reference, but the emission point, where the detected light was emitted, varies from one frame to the next depending on the relative motion of the source to the observer in those frames of reference.

The displacement of the position
Would be when the observer in orbit finds the source is displaced from its predicted position. The opposite of that would be aberration which in that case the observer moving to one side of the source would feel that the source appear to be the other side back whereas she is seen displaced in the opposite position.
This effect which causes by a direct relation between the speed of source and the moment of the observer considered to be of relativistic nature. The angle of aberration, the relative velocity along with a time-based approach might establish a relation that can be applied to the tangential velocity of the observer moment. Incoming light from the source would be shifted at angle determined by the velocity of the orbit. Relative velocity of the radiating light source as to the observer point would be given by diagonal of the velocity vector.

The angle of aberration
$ß$ is considered to be the angle of aberration of radiative light source relative to the moment of the observer. Incoming light from the source would be shifted at angle determined by the velocity of the orbit. The velocity of the radiation along with its direction, angle of aberration relative to the observer formed

by the incoming source velocity constitute the equation $\tan\text{ß}=V\cos\alpha/C-V\sin\alpha$; implying the observer moving at right angles to the source experiences a relative velocity of the radiation from that. When the observer and the source are separated by a distance but both experiencing the same velocity at the same frame of the reference, the travel time for the radiation would be in a simple relationship; $t=\text{distance}/c$, since the observer velocity relative to the source might be ignored in this frame. What is important is when the observer does not move across a line from him/her to the source; in that case the radiation reaches the observer at the speed of the light c with no aberration and the travel time would be same as above; $t=d/c$. If the radiation did not travel direct from the source to the observer in a frame of reference where both are stationary, irrespective of the moment of the frame then the laser adjustment accuracy achieved on earth would not be likely to control as the beam would continue moving relative to the target at a variable rate as to the rotation and orbit of the earth.

Observer perceived cycle time

The source velocity has the effect to raise or lower the energy level of the signal by increase or reduction of the radiation via the change in wavelength, while the radiation velocity remains constant. When the observer moves he/she might not be able to see the radiation from the source in some other frame of reference as being at a constant value of c with regard to the observer, unless the source begin to move at the same velocity as the observer, that is to say, in line with him/her or both would remain stationary at the same reference frame. From having the same velocity as the source every motion from the observer's frame would result in a relative change of the value of the velocity of the radiation. From a moving or stationary within the observer's frame of reference the radiation travels toward the observer at a constant speed. If the observer moves within this frame of reference the result would be a relative change of the value of the the radiation velocity with regard to the observer.

Rotational sources

Rotating radiative sources moving in a plane line with observer would accelerate or decelerate with respect to the observer. The addition of the source velocity to the speed of the radiation ruled out based on the observations by leading scientist Willem de Sitter of the regular motion of twin stars. The result was that the change of speed would not be added to the value of the radiation velocity. What would happen was that the wavelength perceived by the observer would change with an accompanied change in the frequency at a constant value. If the source rotation centre or the observer were also moving, there would be a further overall change of wavelength or cycle time. The resulting effect was that orbiting or rotational radiative bodies would provide means of evaluating their velocity relative to the observer in case the source character was known. The observed frequency would varies to give a peak at max relative rotational velocity.

Low Density Inflationary Universes

We know that the universe is not empty but filled with matter, and ordinary matter through gravity attracts other matter, causing the expansion of the universe to slow down. If the density of the universe exceeds a certain threshold known as the critical density, this gravitational attraction is strong enough to stop and later reverse the expansion of the universe, causing it eventually to recollapse in what is known as the "Big Crunch." On the other hand, if the average density of the Universe falls short of the critical density, the universe expands forever, and after a certain point the expansion proceeds much as if the universe were empty. A critical universe lies precariously balanced between these two possibilities.

Why a Universe of Critical Density?

For quite some time it has been known that the mean density of our universe agrees with the critical density to within better than a factor of ten. Even with such large margin of error this agreement is remarkable. Establishing initial conditions so that the mean density remains close to the critical density for more than a fleeting moment is much like trying to balance a pencil on its point. A universe initially with slightly subcritical density rapidly becomes increasingly subcritical and soon indistinguishable from an empty universe. Similarly, an ever so slightly supercritical universe rapidly collapses into a Big Crunch, never reaching the old age of our universe. For a long time it was regarded simplest and aesthetically most pleasing to postulate that our universe is now of exactly critical density.

The versions of inflation developed in the early 1980s provided a mechanism for setting the density of the universe near the critical density with nearly unlimited precision. For many years an exactly critical universe was touted as one of the few firm predictions of inflation.

Geometry and the Density of the Universe
In Einstein's General Theory of Relativity, formulated in 1915, gravity is understood in terms of geometry rather than as just another ordinary force. Matter tells space-time how to curve and the resulting space-time curvature tells bodies how to move. For the special case of an expanding universe, idealized as filled with a uniform density of matter, a good approximation on large scales, General Relativity establishes an intimate connection between the density of the universe in comparison with the critical density and its geometry.

A universe of critical density (at constant cosmic time) has the familiar Euclidean geometry so well known to us from every experience and from classical perspective as taught in art class. However, a universe of subcritical or supercritical density has a non-Euclidean geometry-hyperbolic if the density is subcritical, or spherical if the density is supercritical. If the curvature of the universe would become apparent only on scales beyond several million light years we might be deceived into believing that its geometry is Euclidean, Only on large scales-larger than the so-called curvature scale--do the differences between the geometries become large effects. The following three plates illustrate the difference in perspective between the three possible geometries: a hyperbolic geometry, a Euclidean geometry, and a spherical geometry. In all three cases, space is divided into identical compartments, whose edges are indicated by the rods. The balls within the compartments are of identical size, and increasing distance is indicated by reddening. In the Euclidean geometry space is divided into cubes and one experiences the ordinary, familiar perspective: the apparent angular size of objects is proportional to the inverse of their distance. Hyperbolic space shown here is tiled with regular dodecahedra. In Euclidean space such a regular tiling is impossible. The size of the compartments is of the same order as the curvature scale. Although perspective for nearby objects in hyperbolic space is very nearly identical to Euclidean space, the apparent angular size of distant objects falls off rapidly. The geometry of spherical space resembles the surface of the earth except here a three dimensional rather than two-dimensional sphere is being considered. Perspective in spherical space is peculiar. Increasingly distant objects first become smaller as in Euclidean space, reach a minimum size, and finally become larger with increasing distance, This behavior could be due to the focusing nature of the spherical geometry.

Stuart Levy of the University of Illinois, Urbana-Champaign and by Tamara Munzer of Stanford University for Scientific American. Copyrighted and reprinted with permission.

What is the Geometry of Our Universe?

During the 1981s observations remained sufficiently crude so that a universe of critical density was quite plausible. But more recent observations have made it increasingly difficult to reconcile a critical universe with the observations. It is known that in addition to the luminous matter seen in the form of stars the universe contains a large amount of "dark" matter, in particular in the halos around galaxies. The presence of this dark matter is inferred from its gravitational pull on the surrounding matter. Since the dark matter is distributed in a less clustered manner than the luminous matter, the apparent average density seems to increase as larger and larger scales are probed. For a long time it was hoped probing sufficiently large scales would uncover a critical density of dark matter. Today it seems unlikely that this hope will ever be realized. It is now possible to probe the average density of the universe on scales large enough to compromise a fair sample of the universe. We present the so-called "cluster baryon fraction" as one illustrative example of the strong evidence in favor of a universe of subcritical density. Rich clusters of galaxies are the largest bound systems in the universe.

Using nuclear physics can determine the baryon density of the universe. With the density of baryonic matter known, the total density can be determined from measuring the baryon fraction. The baryonic mass of a cluster can be determined by adding the masses of the constituent galaxies inferred from their light to the mass of the hot intracluster gas, which can be determined from X-ray observations of emission from the gas. The total mass can be determined by a variety of methods. The motions of the constituent galaxies allow one to determine the depth of the potential well and hence the total mass of the cluster. X-ray observations allow the same to be done with the gas, and gravitational lensing of background objects by the gravitational field of the cluster, resulting in the distortion in appearance of background galaxies, provides a completely independent check of the total mass. These techniques, and a number of independent techniques as well, suggest a universe with approximately one third of the critical density. Although a universe of critical density cannot yet be ruled out definitively, the possibility of a critical universe now appears like quite a long shot.

Reconciling a Low Density Universe with Inflation
If the universe is in fact of subcritical density, does this require abandoning inflation?
If a flat universe really is a "prediction" of inflation as once claimed, one would have to give up inflation. There however exists an escape from this dilemma. Inflation within a single bubble can create a smooth universe with a hyperbolic geometry, just as is required for a universe of subcritical density. Inflation smooths the universe by postulating an early epoch of extremely rapid expansion during which whatever irregularities may have existed prior to inflation are virtually erased. In ordinary inflation, as developed by Guth, Linde, Albrecht, and Steinhardt, this smoothing flattens the universe as well, yielding a universe of critical density. In ordinary inflation, a critical universe could in principle be avoided by shortening the amount of inflation, but in that case the

smoothness on large scales remains a mystery, causing inflation to lose most of its appeal.

The Creation of a Single Bubble Open Universe

In single bubble open inflation there are two epochs of inflation. In inflation the rate of expansion is controlled by a scalar field, known as the inflation field. The inflation field wants to roll down the hill to the bottom and as the field descends the rate of expansion of the universe decreases, eventually ending the epoch of inflationary expansion. In open inflation the inflation field at first remains stuck in a local minimum of the potential. While the field is stuck there, a first epoch of inflationary expansion takes place during which the universe is smoothed. In fact during this epoch the symmetry of the space-time is so large that no particular time direction is preferred over any other. According to classical physics, once stuck in the local minimum the inflation field never escapes; However, quantum mechanics allows the field to tunnel through the barrier, This tunneling occurs through the nucleation of a bubble that subsequently expands.

Subsequently, the bubble expands at the speed of light. It cannot have any velocity other than the speed of light, for else a preferred time direction would be required to exist. The surfaces on the bubble interior on which the scalar field is constant have a hyperbolic spatial geometry, and these are the surfaces that we inside the bubble later perceive as surfaces of constant cosmic time. As one passes inside the bubble, the interior continues to inflate, creating a universe with a large curvature radius. Further inside the bubble the energy of the inflation field is converted into ordinary matter and radiation, and the hyperbolic universe continues to expand and cool down. The best hope for testing open inflation derives from measuring the geometry of the universe, which can be determined through observing the ripples in the cosmic microwave background radiation. The 3K cosmic microwave background radiation emanates from an epoch approximately three hundred thousand years after the Big Bang, when the universe was approximately one thousandth its present size.

At this time the electrons, because of the cooling of the universe, combined with protons and other nuclei to form neutral hydrogen and other elements. Because of this change in composition from a highly ionized plasma to a neutral gas, the formerly opaque universe becomes virtually transparent.

The non-uniformities in the microwave background provide a snapshot of the ripples at that time, which later developed into galaxies and the structure that we observe today. Inflation in general, and open inflation on scales much shorter than the curvature scale, imprints essentially scale free fluctuations on the matter filling the universe. At recombination, however, the physics at that time, believed to be well understood, introduces a preferred scale of known length on which the first acoustic oscillations of the plasma occur. This scale is of known physical size, and from its angle subtended in the sky today, we can determine the geometry of the universe.

M. Bucher and D. Spergel, "Inflation in a Low Density Universe," Scientific American,1999

Light in contact with matter

The theory of interaction of light with matter is called QED. The subject is made to appear more difficult than it is indeed by the very many equivalent methods by which it might be formulated. One of the simplest is that of Fermi. We shall take another starting point by just postulating for the emission or absorption of photons. The transition to quantum electrodynamics involves the assumption that the oscillators are quantum mechanical.

Feynman electrodynamics

In Feynman's words, the field is introduced in the Hamilton picture as a device to hold the track of all those photons that the electron might scatter in the future. You need to know their positions and momenta at the present instant, and you need to determine their future positions by solving their equations of motion along with the electron's. To Feynman, this was an unnecessary package to have to carry along as we travel through space-time. He proposed instead to take an overall space-time view that not only eliminates the need for fields, but is intrinsically relativistic. Like Schwinger and Tomonaga, he recognized the inherent non relativistic nature of the Hamiltonian method. But whereas they stayed within that framework and instead reformulated it by replacing absolute time as the running parameter with one that was suitably relativistic, He proposed to discard the whole method.

Pair Production

If a photon enters matter with an energy in excess of 1.022 MeV it might interact in a procedure called pair production. The photon passing near the nucleus is subjected to strong field effects and might disappear as a photon and reappear as a positive negative electron pair. The two electrons produced e-,e+ are not scattered orbital electrons but are created, de novo, in the energy/mass conversion of the disappearing photon. In Pair Production, the energy of the photon is converted into mass, and the remnant of energy not absorbed as mass energy is converted into kinetic energy. Thus, the correlation of the DP associated with the photon is fully converted to the particle- centered form of energy. The photon travels only at the speed of light, whereas the mass travels at a rate that reflects the amount of energy associated with the amount of kinetic energy remaining after forming the mass. Conventional physics recognizes that gamma rays passing close to an atomic nucleus form an electron-positron pair. But the conventional physics model of the universe does not have a robust mechanism for offering how a photon traveling through the vacuum of space could produce such an effect. Two possible mechanisms for pair production might be at place: Either the pair decays from the g-ray itself as to the fact that gamma ray is an electron-positron pair that splits into a pair in an environment such as the stressed space of the nucleus, Or Space is filled with electron-positron pairs which can be re-energized into real particles provided the heavy nucleus and properly oriented high energy photon. The Theory is a study in the mechanisms observed in nature, and a subsequent examination of the consistency of the theoretical mechanism with the experimental evidence. In the case of pair production we see a mechanism by which particles transform between one type of Dipole to another. Such transformations between various states of energy and interactions between various configurations of energy are the essence of the processes of life. The conversion of the energy of the photon into the energy of the positron and electron mass takes work. The original state of the Dipole was a flat uniformly distributed volume of equidistant particles. By adding it into Dipole and modifying it through a special process it takes the configuration of a mass. The order of Dipole held by Photon is complete and conserved as its order is transformed into a new type of order as the electron and positron. The key principle which produced the conversion of the high energy gamma ray into the positron and electron is the longitudinal stretching of the inner and outer limb of the g ray photon. Such a process is the modifying of the very singular nature of photon that its integrity can be severed into two parts while positron and electron can recombine. In particular, the 2nd law of thermodynamics, which states that the interactions always increase or maintain the amount of randomness in a system. Entropy is the measure of randomness in a system, and clearly the breaking of a 1.022 Mev photon into two .511 Mev photons is an increase in the entropy of the system. Entropy is not the driving factor that moves particles, and it does not cause or prevent transformations of state.

Entropy includes the concept that a system has a particular state, that a force can disturb the state, and that the particles or elements of the system can be moved away from that state of order. In the case of pair production and the observed photon split, a force was applied to the photon which overcame the strength of its internal boundaries. By stretching the photon's wave-front beyond its ability to continue to transform the E field into B field, and vice versa, the photon's internal integrity was broken. The Dipole used to be the place for the photon at which it separates into two and thereby expand its randomness.

Such is the general consideration in all cases of transformation of structure and state; a force is applied to the system which is greater than the force constraining its current configuration. Obviously, transition into a new state cannot take place if there is adequate force against it.

"What are the forces that are acting on the system to change its state"?

"what is the energy barrier opposing the transition"?

The splitting of the photon into two requires an activation energy, Systematic procedures require an activation energy to initiate a reaction that changes state. On physical level, the forming or breaking of a molecular bond requires an activation energy. In the case of bonding together two atoms to form a molecule, the repulsion of the outer atomic shells must be overcome with the kinetic energy that is directing the two atoms toward collision. In effect, the energy contained within the system will be used as the force acting on the two atoms to create a new state. The collision of two atoms deforms the outer orbitals, enables them to be in a position where it is favorable to share electrons, and thus forming a new molecular species. The kinetic energy contained in the relative motion of the two particles enabled them to overcome the barrier state of their mutual outer orbital repulsion. Returning to the examination of a photon splitting into an electron and positron, the activating energy of this system is provided by the photon which has sufficient Electromagnetic energy to stored within its structure to form two constituent particles. The photon should have been entered into a space where it can undergo a reaction with an environment . The velocity of the photon which moved it toward the nucleus, the organizational state and internal dynamic process of the photon, and the gradient of conduction density provided by the nucleus, all came together to provide an environment where the internal structure of the photon was broken apart to form two particles in the place of a single photon. The high-energy g ray photon splits into two regions due to the differential in speed of light between the photon's inner and outer limb. The Dipole is separated into a positive and negative region by the E field of the photon. The balance of the photon's EM energy not converted into mass-energy is converted into kinetic energy or photonic energy.

The stretching of the photon between the inner and outer limb, and the precipitation of the electron and positron into these two regions consumes a portion of the photon's E field energy. But, if the energy of the photon is greater than 1.022 Mev, then the rest of the energy of the photon should either transferred to another photon, or converted into kinetic energy associated with the masses. The method by which the photon converts its energy into kinetic energy is as follows:

1 the E field splits the Dipole into a positron and electron,

2 The remnant of the E field not absorbed as mass-energy then acts on the electron and positron and forces them in opposite directions.

The E field that splits the DP into a positron and electron was initially operating as a field with a particular orientation. The concept of the E field having a direction means that positive charges move in one direction under its influence, and negative charges move in the other. Thus, as the photon breaks into two separate regions, the E field will still be acting on the newly created electron and positron to accelerate them. And since an E field, which has only a single direction, will push a positive and negative particle in opposite directions, the remnant of the photon's E field will accelerate the new positron and electron and cause them to separate. The heavier the nucleus, the higher the gradient of negative and positive DP formed in the volume close to the surface of the nucleus. The higher concentration of positive and negative DPs around a heavier nucleus means that there is larger volume available for possible pair production, and hence a greater probability of a gamma ray producing an electron positron pair.

This is why a heavy nucleus such as lead is used as a radiation shield for rays. A heavy nucleus is more effective than lighter elements in causing Pair Production. After the pairs are formed, they cancel out each other, creating lower energy gamma rays that may cause further pair production or ionization. Electrons will be recaptured, and vibrate the lattice. The pathway is in essence a downward path of energy concentration from a high energy gamma to low energy thermal vibration of many molecules. A quantum concept is the fact that the G-ray should provide at least the 1.022 MeV of energy to generate sufficient energy to form the organizational charge structure of the electron and positron. This amount of organizational energy is needed to separate out a negative and positive from the DP and get them far enough apart that they can exist as separate entities. The G-ray and all photons, have an alternating E field and B field, and the conventional symmetry arguments refer to this as angular momentum, or spin; And, after electron and positron are formed, they both are considered to have a spin. Evidence of this spin, is seen in the Pauli Exclusion Principle which does not allow two electrons in the same electron orbital to have the same spin.

The Stern-Gerlach experiment
From the time of Ampere onward, molecular currents were regarded as giving rise to magnetic moments. In the nuclear model of the atom the electron orbits the nucleus. This circular current results in a magnetic moment. The atom behaves as if it were a tiny magnet. In the Stern-Gerlach experiment a beam of silver atoms passed through an in-homogeneous magnetic field. In Larmor's classical theory there was no preferential direction for the direction of the magnetic moment and so one predicted

that the beam of silver atoms would show a maximum in the center of the beam. In Sommerfeld's quantum theory an atom in a state with angular momentum equal to one/L=1, would have a magnetic moment with two components relative to the direction of the magnetic field, $\pm$e /4m.

In an in-homogeneous magnetic field, H, the force on the magnetic moment will be zx (Gradient of the magnetic field in the z direction), where z = $\pm$ e /4m, where e is the charge of the electron, m is its mass, h the Planck's constant, and z is the field direction. Thus, depending on the orientation of the magnetic moment relative to the magnetic field there will be either an attractive or repulsive force and the beam will split into two components, exhibiting spatial quantization. According to quantum theory z can only be $\pm$ (e/2me)(/2π).

In this case the spot on the receiving plate will therefore be split into two, each of them having the same size but half the intensity of the original spot. This difference in prediction between the Larmor and Sommerfeld theories was what Stern and Gerlach planned to use to distinguish between the two theories. Stern remarked that the experiment, if it can be carried out, will result in a clear-cut decision between the quantum-theoretical and the classical. Sommerfeld's theory also acted as an enabling theory for the experiment. It provided an estimate of the size of the magnetic moment of the atoms so that Stern could begin calculations to see if the experiment was feasible. Stern calculated, for instance, that a magnetic field gradient of 104 Gauss per centimeter would be sufficient to produce deflections that would give detectable separations of the beam components. He asked Gerlach if he could produce such a gradient. Gerlach responded affirmatively, and said he could do even better. The experiment seemed feasible. The silver atoms pass through the inhomogeneous magnetic field. If the beam is spatially quantized, as Sommerfeld predicted, two spots should be observed on the screen. The beam splitting into three components, which would be expected in modern quantum theory for an atom with angular momentum equal to one. A preliminary result reported by Stern and Gerlach did not show splitting of the beam into components.

It did, however, show a broadened beam spot. They concluded that although they had not demonstrated spatial quantization, they had provided evidence that the silver atom possesses a magnetic moment. Stern and Gerlach made improvements in the apparatus, particularly in replacing a round beam slit by a rectangular one that gave a much higher intensity. There was an intensity minimum in the center of the pattern, and the separation of the beam into two components was clearly seen. This result seemed to confirm Sommerfeld's quantum-theoretical prediction of spatial quantization. The Stern-Gerlach result posed a problem for the Bohr-Sommerfeld theory of the atom.

Stern and Gerlach had assumed that the silver atoms were in an angular momentum state with angular momentum equal to one ($L = 1$). In fact, the atoms are in an $L = 0$ state, for which no splitting of the beam would be expected in either the classical or the quantum theory. The later, or new, quantum theory developed by Heisenberg, Schrödinger and others predicted that for L=1 state the beam should split into three components. The magnetic moment of the atom would be either 0 or $\pm$ e / (4π x m); Thus, if the silver atoms were in L = 1 state as Stern and Gerlach had assumed, their result, showing two beam components, also posed a problem for the new quantum theory. This was solved when Uhlenbeck and Goudsmit (1925, 1926) proposed that the electron had an intrinsic angular momentum or spin equal to /4π. This is analogous to the earth having orbital angular momentum about the sun and also an intrinsic angular momentum due to its rotation on its own axis. In an atom the electron will have a total angular momentum J = L + S, where L is the orbital angular momentum and S is the spin of the electron. For silver atoms in an L = 0 state the electron would have only its spin angular momentum and one would expect the beam to split into two components. Goudsmit and Uhlenbeck suggested the idea of electron spin to explain features in atomic spectra such as the anomalous Zeeman effect, the splitting of spectral lines in a magnetic field into more components than could be accommodated by the Bohr-Sommerfeld theory of the atom. Although the Stern-Gerlach results were known, and would certainly have provided strong support for the idea of electron spin, Goudsmit and Uhlenbeck made no mention of that.

The Stern-Gerlach experiment was initially regarded as a crucial test between the classical theory of the atom and the Bohr-Sommerfeld theory. In a sense it was, because it showed clearly that spatial quantization existed, a phenomenon that could be accommodated only within a quantum mechanical theory. It decided between the two classes of theories, the classical and the quantum mechanical. With respect to the particular quantum theory of Bohr and Sommerfeld, however, it wasn't crucial, although it was regarded as such at the time, because that theory predicted no splitting for a beam of silver atoms in the ground state. The theory had been wrongly applied. The two-component result was also problematic for the new quantum theory, which also predicts no splitting for an angular momentum zero state and three components for L = 1 state.

Particle scattering and cross sections

When particles interact with a target; most of them continue on unscattered, some of them interact with the target and scatter. Those that do scatter do so at a particular angle in three dimensions, i.e. you give the scattering angle as a solid angle $d\Omega$ which equals $\sin\theta \, d\theta d\phi$, where ϕ and θ are the spherical angles.

The number of particles scattered into a specific $d\Omega$ per unit time is proportional to a very important quantity in scattering theory which is the differential cross section given by $d\delta(\phi,\theta(/d\Omega$ as the measure of the number of particles per second scattered into $d\Omega$ per incoming flux. The incident flux , J also called the current density is the number of incident particles per unit area per unit time.

Photons splitting in two

Whatever happened to one particle would thus immediately

affect the other particle, wherever in the universe it may be. Einstein called this "Spooky action at a distance." When photon splits into two photons, the resulting photon pair is considered entangled. Researchers have detected glimpses of a rare event in which a single photon splits in two. If a fat man walked into an empty room and then two skinny guys walked out, you might be perplexed. Now physicists have spotted the equivalent result in photons flying near an atom. A group publishing in the 5 August 2002 print issue of PRL has identified rare instances in which a single photon splits in two, dividing the original photon's energy between them. Fundamental particles constantly and randomly morph into "virtual" particles. A photon, for example, can temporarily become an electron and positron which quickly cancel out each other to reform the original photon. In a vacuum the process has little effect, but the electric field of an atom can interact with electron-positron pairs to create theoretic measurable results, some of which have already been fingered. One such event, which researchers had sought for decades, is photon splitting.

In unrelated experiments, physicists studying quantum optics often create pairs of photons from single, higher energy photons by hitting a special crystal with laser light, but that process involves exciting the crystal's atoms. In photon splitting, a photon first transforms into an electron-positron pair; then one of those particles emits a photon before cancel out its partner to produce the second photon. In 1995, a team at the Budker Institute of Nuclear Physics BINP in Novosibirsk, made the first observation of photon splitting, and they announced preliminary results at two conferences. They completed their full analysis of the data, which turned up more splitting events than before and allowed them to compare the predictions of exact quantum field theory with the conventional approximation. The team used highly energetic gamma rays; photon energies between 100 and 450 MeV, which they produced by colliding infrared photons head-on with a high-energy electron beam. These pumped up photons were less likely to engage in interactions that would have obscured detection of the split photons. The high energy beam passed through a bismuth germanate crystal target into detectors, of 1.6 billion photon hits, only around 405 photon pairs fit the bill.

The Stern-Gerlach Experiment and Spin-1/2 Systems

The Original Experiment In 1922 Otto Stern and Walther Gerlach sent a beam of silver atoms (a spin-1/2 system) through an inhomogeneous magnetic field in the z direction. Since the silver atoms have an intrinsic magnetic moment (or spin angular momentum) they should be deflected by the inhomogeneous magnetic field in the z direction depending on their orientation with respect to the magnetic field. When Stern and Gerlach performed the experiment, they expected a uniform spread of the beam in the z direction to result. Much to their surprise, the inhomogeneous magnetic field effectively split the beam in to two parts. This result lead to the idea of quantization of spin angular momentum such that the component of spin in a particular direction can only take on two values: +h/2 or -h /2, or what we also call spin up and spin down, even though the spin itself is not up or down, it is the particular direction's component that is up or down.

The Ideal Stern-Gerlach Apparatus

These simulations make use of two types of ideal Stern-Gerlach apparatus to spatially separate the spin-1/2 particles: those that use a transverse magnetic field and those that use a longitudinal magnetic field. Both are ideal in the sense that there is no experimental error associated with using them. In other words, the outcomes are always exactly those predicted by quantum theory. Because of this idealization, we do not distinguish between the apparatus internal mechanisms; they are all simply referred to as ideal Stern-Gerlach apparatus. The incident beam can either be a beam of a random or statistical mixture of spin orientations, abeam of a particular spin eigenstate such as |z+> or |z − >, or a beam of a particular superposition of eigenstates (such as 0.707 [z+> + |z − >]). Once through one or more ideal Stern-Gerlach apparatus, the output is detected at the counters represented by the horizontal bars and their associated numbers.

Transverse Stern-Gerlach effect

The original experiment by Stern and Gerlach made use of a transverse in-homogeneous magnetic field generated by a permanent magnet perpendicular to the direction of propagation of the spin-1/2 particles. So if a particle is moving in the x direction and is subject to an in-homogeneous field in the z direction, it will experience a force, $F_z \sim \mu_z dB_z/dz$. Such a force will deflect the beam either up or down due to the fact that $\mu z = gqSz/2mc$, where g is the gyromagnetic ratio, q is the particle's charge, m is the mass of the particle, and Sz is the z component of the particle's spin. Therefore such an apparatus can be used to spatial separate particles whose z component of spin are oriented either up or down.

Longitudinal Stern-Gerlach effect

These simulations also show a different type of ideal Stern-Gerlach apparatus: one which measures spin in the direction of propagation using, in principle, the longitudinal SternGerlach effect. If we again assume that the direction of propagation is the x direction, the ideal Stern-Gerlach apparatus must have an in-homogeneous magnetic field in this direction. It is difficult to see how this can occur with a permanent magnet, but not too difficult to imagine with magnetic field generated by a current in a coil that the beam of spin-1/2 particles passes . The spin-1/2 articles are then "deflected" either forward or backward thereby generating a spatial separation that in principle could be detected.

Cavity Quantum Electrodynamics

Cavity QED investigates the interaction of single atoms with single electromagnetic field modes. To achieve the experimental goal of realizing such a system, effort was made and nowadays it is achievable even for optical transitions. This paves the way for many interesting physical applications. Two of the most interesting ones are for sure the use of cavity QED for the construction of a quantum network and on the other hand its usefulness for elementary verifications of quantum mechanics.

Fabian Grusdt

Cavity

The cavity is a basic concept in all continuum models. The model in fact is composed of an atom or a few atoms, put into a void cavity within a continuous dielectric medium. The shape and size of the cavity are differently defined in the various versions of the continuum models. As a general rule, a cavity should have a physical meaning, such as that introduced by Onsager, and not be only a mathematical artifice as often happens in other descriptions of such effects. In particular, the cavity should contain within its boundaries the largest likely part of the charge distribution. These requirements are in contrast with the description of the whole system given by any QM level. The electronic charge distribution of an isolated molecule; in fact, persists to infinity. In a condensed medium the conditions at large distances are less well-defined, but in any case there will be an overlap with the charge distribution of the medium, not explicitly described in continuum models but existing in real systems. In continuum models, much attention has been paid to the portion of medium electric charge outside the boundaries of the cavity; the terms "escaped charge" and "outlying charge" are often used to indicate this portion of charge. Here we will assume that all of the charge distribution lies inside the cavity, which in turn has a size not so large as to be in contrast with the medium exclusion postulate. The optimal size of the cavity has thus been a subject of debate, and several definitions have been proposed. The adopted definitions are the result of a trade-off between conflicting physical requirements. The shape of the cavity has also been the object of many proposals. It is accepted that the cavity shape should reproduce as well as possible the atomic shape. Cavities not respecting this condition may lead to deformations in the charge distribution after the medium polarization. Here, once again, there is a trade-off between computational exigencies and the desire for better accuracy. Quantum mechanical calculations of the molecular surface can give a direct ab initio definition of the cavity. An accurate description is based on the use of a surface of constant electronic density. Within this framework, one only needs to specify the isodensity level in the cavity environment.

Photon Cavity

Serge Haroche main research activities have been in quantum optics and quantum information science. He has made important contributions to Cavity Quantum Electrodynamics CQED, the domain of quantum optics which studies the behaviour of atoms interacting strongly with the field confined in a high-Q cavity. An atom-photon system isolated from the outside world by highly reflecting metallic walls realizes a very simple experimental model which Serge Haroche has used to test fundamental aspects of quantum physics such as state superposition, entanglement, complementarity and decoherence. Some of these experiments are actual realizations in the laboratory of the "thought experiments" imagined by the founding fathers of quantum mechanics. Serge Haroche's main achievements in cavity QED include the observation of single atom spontaneous emission enhancement in a cavity 1983, the direct monitoring of the decoherence of mesoscopic superpositions of states so-called Schrödinger cat states 1996 and the quantum-nondemolition measurement of a single photon 1999. By manipulating atoms and photons in high-Q cavities, he has also demonstrated many steps of quantum information procedure such as the generation of atom/atom and atom/photon entanglement 1997, the realization of a photonic memory 1997 and the operation of quantum logic gates involving photons and atoms as quantum bits known as Qubits 1999.

The basic ingredients of CQED

Two-level systems qubits+quantum harmonic oscillators play an important role in this physics. The qubits are information carriers and the oscillators act as memories or quantum bus linking the qubits together; Coupling qubits to oscillators is the domain of Cavity Quantum Electrodynamics.

CQED

Atoms and photons in small cavities behave completely unlike those in free space. Their quirks illustrate some of the principles of quantum physics and make possible the development of new sensors. Fleeting spontaneous transitions are ubiquitous in the quantum world. Once they are under way, they seem as uncontrollable and as irreversible as the explosion of fireworks. Excited atoms, for instance, discharge their excess energy in the form of photons that escape to infinity at the speed of light. Yet during the past decade, this inevitability has begun to yield. Atomic physicists have created devices that can slow spontaneous transitions, halt them, accelerate them or even reverse them entirely. Recent advances in the fabrication of small superconducting cavities and other microscopic structures as well as novel techniques for laser manipulation of atoms make such feats possible. By placing an atom in a small box with reflecting walls that constrain the wavelength of any photons it emits or absorbs- and thus the changes in state that it might undergo-investigators can cause single atoms to emit photons ahead of schedule, stay in an excited state indefinitely or block the passage of a laser beam.

With further refinement of this technology, cavity quantum electrodynamics should find use in the generation and precise measurement of electromagnetic fields consisting of only a handful of photons. Cavity QED processes engender an intimate correlation between the states of the atom and those of the field, and so their study provides new insights into quantum aspects of the interaction between light and matter. To understand the interaction between an excited atom and a cavity, one should keep in mind two kind of physics: the classical and the quantum. The emission of light by an atom bridges both worlds, light waves are moving oscillations of electric and magnetic fields. In this respect, they represent a classical event, however light can also be described in terms of photons, discretely emitted quanta of energy. Sometimes the classical model might be the best and from time to time the quantum one offers better understanding. When an electron in an atom jumps from a high energy level to a lower one, the atom emits a photon that carries away the difference in energy between the two levels. This photon has wavelength of a micron or less, corresponding to a frequency of a few hundred tetra-hertz and an energy of about one electron volt. Any given excited state has a natural lifetime similar to the half-life of a radioactive element_that determines the odds that the excited atom will emit a photon during a given time interval. The probability that an atom will remain excited decreases along an exponential curve to one half after one tick of the internal clock, one quarter after two ticks an so on. In classical terms, the outermost electron in an excited atom is the equivalent of a small antenna, oscillating at frequencies corresponding to the energy of transitions to less excited states, and the photon is simply the antenna's radiated field. When an atom absorbs light and jumps to a higher energy level, it acts as a receiving antenna instead.

If the antenna is inside a reflecting cavity, however, its behaviour changes as anyone knows who has tried to listen to a radio broadcast while driving through a tunnel. As the car and its receiving antenna pass underground, they enter a region where the long wave-lengths of the radio waves interfere destructively with those that bounce off the steel-reinforced concrete walls of the tunnel. In fact, the radio waves cannot propagate unless the tunnel walls are separated by more than half a wave-length. This is the minimal width that permits a standing wave with at least one crest, or field maximum, to build up-just as the vibration of a violin string reaches a maximum at the ends. What is true for reception also holds for emission: a confined antenna cannot broadcast at long wavelengths. An excited atom in a small cavity is precisely such an antenna, albeit a microscopic one. If the cavity is small enough, the atom will be unable to radiate because the wavelength of the oscillating field it would like to produce cannot fit within the boundaries. As long as the atom cannot emit a photon, it has to remain in the same energy level; excited state acquires an infinite lifetime. In 1985 research groups at the University of Washington and at the Massachusetts Institutes of Technology demonstrated suppressed emission.

The group in seattle inhibited the radiation of a single electron inside an electromagnetic trap, whereas the MIT group studied excited atoms confined between two metal plates about a quarter of a millimeter apart. The atoms remained in the same state without radiating as long as they were between the plates. Millimeter-scale structures are much too wide to alter the behaviour of excited atoms emitting micron or sub-micron radiation; consequently, the MIT experiments had to work with atoms in special states known as Rydberg state has almost enough energy to lose an electron completely. Because this outermost electron is bound only weakly, it can assume any of a great number of closely spaced energy levels, and the photons it emits while jumping from one to another have wavelengths ranging from a fraction of a millimeter to a few centimeters. Rydberg atoms are prepared by irradiating ground-state atoms with laser light of appropriate wavelengths and are widely used in cavity QED experiments. The suppression of spontaneous emission at an optical frequency requires much smaller cavities.

In 1986 one of us; Haroche, along with other physicists at Yale University, made a micron-wide structure by stacking two optical flat mirrors separated by extremely thin metal spacers. The workers sent atoms through, thereby preventing them from radiating for as long as 13 times the normal excited-state lifetime. Researchers at the University of Rome used similar micron-wide gaps to inhibit emission by excited dye molecules. The experiments performed on atoms between two flat mirrors have an interesting twist. Such a structure, with no sidewalls, constrains the wavelength only of photons whose polarization is parallel to the mirrors.

As a result, emission is inhibited only if the atomic dipole antenna oscillates along the plane of mirrors. The Yale researchers demonstrated these polarization-dependent effects by rotating the atomic dipole between the mirrors with the help of a magnetic field. When the dipole orientation was tilted with respect to the mirrors' plane, the excited state lifetime dropped. Suppressed emission also takes place in solid-state cavities-tiny regions of semiconductor bounded by layers of disparate substances. Solid-state physicists routinely produce structures of sub-micron dimensions by means of molecular-beam epitaxy, in which materials are built up one atomic layer at a time. Devices built to take advantage of cavity QED could engender a new generation of light emitters. These experiments indicate a counter-intuitive phenomenon that might be called " no-photon interference". In short, the cavity prevents an atom from emitting a photon because that photon would have interfered destructively with itself had it ever existed; however this implies that the photon know even before being emitted whether the cavity is the right or wrong size. Part of this might be due to the odd result of quantum mechanics. A Cavity with no photon is in its lowest energy state, the so-called ground state, though not empty. The Heisenberg uncertainty sets a lower limit on the product of the electric and magnetic fields inside the cavity or anywhere else for that matter and thus prevents them from simultaneously vanishing.

This so-called vacuum field exhibits intrinsic fluctuations at all frequencies, from long radio waves down to visible ultraviolet and gamma radiation and is a crucial concept in theoretical physics. Indeed, spontaneous emission of a photon by an excited atom is in a sense induced by vacuum fluctuations. The no-photon interference effect arises because the fluctuations of the vacuum field, like the oscillations of more actual electromagnetic waves are constrained by the cavity walls. In a small box, boundary conditions prevent long wavelengths-there can be no vacuum fluctuations at low frequencies. An excited atom that would ordinarily emit a low-frequency photon cannot do so, because there are no vacuum fluctuations to simulate its emission by oscillating in phase with it. Small cavities suppress atomic transitions; slightly larger ones, however, can enhance them. When the size of a cavity surrounding an excited atom is increased to the point where it matches the wavelength of the photon that the atom would emit, vacuum-field fluctuations at that wavelength flood the cavity and become stronger than they could be the case in free space. This state of affairs encourages emissions; the lifetime of the excited state becomes much shorter. We observed this emission enhancement with Rydberg atoms at the Ecole Normale Superieure ENS in Paris in one of the first cavity QED experiments back in 1983. If the resonant cavity has absorbing walls or allows photons to escape, the emission is not different from spontaneous radiation in free space in its essence_ It just proceeds much faster. If the cavity walls are very good reflectors and the cavity is closed, however, novel effects occur.

These effects, which depend on intimate long-term interactions between the excited atom and the cavity, are the basis for a series of new devices that can make sensitive measurements of quantum phenomena. Instead of simply emitting a photon and going on its way, an excited atom in such a resonant cavity oscillates back and forth between its excited and unexcited states. The emitted photon remains in the box in the vicinity of the atom and is promptly reabsorbed. The atom-cavity system oscillates between two states, one consisting of an excited atom and no photon, and the other of a de-excited atom and a photon trapped in the cavity. The frequency of this oscillation depends mostly on the transition energy, on the size of the atomic dipole and on the size of the cavity. This atom-photon exchange has a deep analogue in classical physics; If two identical pendulums are coupled by a weak spring and one of them is set in motion, the other will soon start swinging while the former comes to rest. At this point, the former starts swinging again, commencing an exchange of energy. A state in which one pendulum is excited and the other is at rest is not stationary, because energy moves continuously from one pendulum to the other. The system does have two steady states, however; one in which the pendulums swing in phase with each other, and the other in which they swing alternatively toward and away from one another.

The system's oscillation in each of these eigenmodes differs due to the additional force imposed by the coupling_the pendulums oscillate slower in phase and faster in out of phase. Furthermore, the magnitude of the frequency difference between the two eigenmodes is equal to the rate at which the two pendulums exchange their energy in the non-stationary states.

Researchers recently observed this mode splitting in an atom-cavity system. They transmitted a weak laser through a cavity made of two spherical mirrors while a beam of cesium atoms also passes through the cavity. The atomic beam was so tenuous that there was at most one atom at a time in the cavity. Although the cavity was not closed, the rate at which it exchanged photons with each atom exceeded the rate at which the atoms emitted photons that escaped the cavity. The spacing between the mirrors was an integral multiple of the wavelength of the transition between the first excited state of cesium and its ground state. Experimenters varied the wavelength and hence the frequency of the laser and recorded its transmission across the cavity. When the cavity was empty, the transmission reached a sharp maximum at the resonant frequency of the cavity. When the resonator contained one atom on average, however, a symmetrical double peak appeared. The frequency splitting about six mega-hertz marked the rate of energy exchange between the atom and a single photon in the cavity. This apparatus is extremely sensitive; when the laser is tuned to the cavity's resonant frequency, the passage of a single atom lowers transmission significantly. This phenomenon can be used to count atoms in the same way one currently counts cars or people intercepting an infrared light in front of a photo-detector.

The cavity must be as small as possible because the frequency splitting is proportional to the vacuum-field amplitude, which is inversely proportional to the square root of the box's volume. At the same time, the mirrors must be very good reflectors so that the photon remains trapped for at least as long as it takes the atom and cavity to exchange a photon. Experimenters have been able to achieve longer storage times-as great as several hundred milliseconds- by means of superconducting niobium cavities cooled to temperatures of about one kelvin. These cavities are ideal for trapping the photons emitted by Rydberg atoms, which typically range in wavelength from a few millimeters to a few centimeters. In a recent experiment in our laboratory at ENS we excited rubidium atoms with lasers and sent them across a superconducting cylindrical cavity tuned to a transition connecting the excited state to another Rydberg level 68 Gigahertz higher in energy; we observed a mode splitting of about 100 K-hertz when the cavity contained three or two atoms at the same time. There is a striking similarity between the single atom-cavity system and a laser-induced. Either device, which emits photons in the optical and microwave domain, respectively, consists of a tuned cavity and an atomic medium that can undergo transitions whose wavelength matches the length of the cavity.

When energy is supplied to the medium, the radiation field inside the cavity builds up to a point where all the excited atoms undergo stimulated emission and give out their photons in phase. Indeed, in 1984 physicists at the Max Planck Institute for quantum optics succeeded in operating a micro-maser containing an atom. To start up the micro-maser, Rydberg atoms are sent one at a time through a superconducting cavity. These atoms are prepared in a state whose favored transition matches the resonant frequency of the cavity. This apparatus is another diversification of the atom-cavity coupled oscillator; if an atom were to remain inside that, it would exchange a photon with the cavity at some characteristic rate. Instead, depending on the atom's speed, there is some fixed chance that an atom will exit unchanged and a complementary chance that it will leave a photon behind. If the cavity remains empty after the first atom, the next one faces an identical chance of exiting, the cavity in the same state in which it entered. However, an atom deposits a photon; then the next atom in line encounters sharply altered odds that it will emit energy. The rate at which atom and field exchange energy depends on the number of photons already present_ the more photons, the faster the atom is stimulated to exchange additional energy with the field. Soon the cavity contains two photons, modifying the odds for subsequent emission even further, then three and so on at a rate that depends at each step on the number of previously deposited photons. In fact, the photon number does not increase without limit as atoms keep passing the resonator. Because the walls are not perfect reflectors, the more photons there are, the greater becomes the chance that one of them will be absorbed.

About hundred thousand atoms per second can come through a typical micro-maser, each remaining perhaps 10 microseconds, meanwhile the photon lifetime within the cavity is about 10 milliseconds. Consequently, such a device running in steady state contains about 1000 microwave photons, Each of them carries an energy of bout .0001 electron-volt. Though it would be difficult to measure such a tiny field directly, the atoms going through the resonator provide a very simple way to monitor the maser. The transition rate from one Rydberg state to the other depends upon the photon number in the cavity, and experiments need only measure the fraction of atoms leaving the maser in each state. The populations of the two levels can be determined by ionizing the atoms in two small detectors, each consisting of plates with an electric field across them. The first detector operates at a low field to ionize atoms in the higher-energy state; the second operates at a slightly higher field to ionize atoms in the lower-state, i.e. those that have left a photon behind. With its tiny radiation output and it drastic operational requirements, the micro-maser may or may not be a machine that could be taken off a shelf and switched on by pushing a knob. It is an ideal system to illustrate and test some of the principles of quantum physics. Intriguing variation of the micro-maser is the two-photon maser source.

Such a device was operated for the first time several years ago by our group; Atoms went through a cavity tuned to half the frequency of a transition between two Rydberg levels. Under the influence of the cavity radiation, each atom was stimulated to emit a pair of identical photons, each bringing half the energy required for the atomic transition, The maser field built up as a result of the emission of successive photon pairs. The presence of an intermediate energy level near the midpoint between the initial and the final levels of the transition helps the two-photon procedure along. Loosely speaking, an atom goes from its initial level to its final one via a virtual transition during which it jumps down to the middle level while emitting the first photon; it then jumps down again while emitting the second photon. The intermediate step is virtual because the energy of the emitted photons, whose frequency is set by the cavity, does not match the energy differences between the intermediate level and either of its neighbors. How can such a paradoxical situation exist? The micro-maser cavity makes two photon operation possible in two ways;

1) It inhibits single-photon transitions that are not resonant with the cavity

2) It strongly enhances the emission of photon pairs

Without the cavity Rydberg atoms in the upper level would radiate a single photon and jump down to the intermediate level. This procedure would deplete the upper level before two photon emission could build up. Though the basic principle of a two photon micro-maser is the same as that of its simple one photon cousin, the way in which it starts up and operates differ significantly. A strong fluctuation, corresponding to the unlikely emission of several photon pairs in close succession might be required to trigger the system; as a result, the field builds up after a period of lethargy. Once this fluctuation has occurred the field in the cavity is relatively strong and stimulates emission by subsequent atoms causing the device to reach full power rapidly. A two-photon laser system recently developed at Oregon state University operates along a different scheme but displays the same metastable behaviour. The success of micro-masers and other similar devices has prompted cavity QED researchers to conceive new experiments, some of which might have been dismissed as pure science fiction only a few years ago. Perhaps the most remarkable of these as yet hypothetical experiments are those that deal with the forces experienced by an atom in a cavity containing only a vacuum or a small field made of a few photons. The first thought experiment starts with a single atom and an empty cavity tuned to a transition between two of the atom's states. This coupled oscillator system has two non-stationary states; One corresponds to an excited atom in an empty cavity; the other to a de-excited atom with one photon.

The system also has two stationary states, obtained by addition or subtraction of the non-stationary ones_addition corresponds to the in-phase oscillation mode of the two pendulum model and subtraction corresponds to the out of phase mode; these stationary states differ in energy by a factor equal to Planck's constant; h, times the exchange frequency between the atom and the field.

Serge Haroche, Jean-Michel Raimond

Qubits and Oscillators_Description of a qubit or spin 1/2

A pure state of a qubit |0>,|1> is parametrized by two polar angles θ,ϕ and is represented by a point on the Bloch sphere. The Bloch vector components are the expectation values of the Pauli operators. The qubit state is determined by performing averages on an ensemble of realizations. Coupling qubits to oscillators is an important ingredient in quantum information.

Coherent state

Coupling field mode to classical source generates coherent state whose amplitude increases linearly with time. A classic antenna would be located at r=0. λ would be a constant proportional to current amplitude in antenna. The rotating wave approximation keeping time independent. For the Field evolution in cavity starting from vacuum at t=0 we use Glauber formula. Expanding in power series, we get the field in Fock state basis ωfield |t>, Coupling field mode to classical source generates coherent state whose amplitude increases linearly with time.

Useful Formula
Density operator for a field mode
ρ=ωfield ωfield /pure state
ρ=Σpi |ω(i(field ω(i(field /mixed state

Matrix elements of ρ defined as the wave function of the field for the pure state and the sum of the complex wave function with an imaginary term for the mixed state; can be discrete in Fock state basis or continuous in quadrature basis. Going from one representation to the other is easy knowing the amplitudes x|n> expressing the oscillator energy eigenstates.

Bayes law or projection postulate
Comes from the work by the Reverend Thomas Bayes is the probability to detect the atom in state j (0 or 1) provided they are n photons in cavity would be the reciprocal probability provided that the atom has been detected in j is given by Bayes law.

Within normalization, the inferred photon number probability is the a priori one multiplied by the Ramsey fringe function. The same result is obtained by applying the projection postulate to the qubit measurement.

Bayes law/Maximum Likelihood estimator

A natural choice for the estimator is maximised by bayes combined with some assumption of prior knowledge about the estimator θ before the approximation. The joint probability for finding values of couple estimators $p(x,\theta$ can be expressed in terms of the a priori probabilities over each of them, i.e. px and $p\theta$, If nothing is a priori known about one of them,let's say; θ, we should assume a flat $p\theta$ probability distribution leading to probabilities of each of them over their intgrated sum; $p(\theta/x = p(x/\theta) / \int p(x/\theta)d\theta$. The probability distribution of one of the estimators;e.g.,θ after result of the other; e.g., x has been found is thus given by the likelihood function $p(x/\theta$.

Photon count in cavity

To count up to nm photons we can either choose $\phi o = \pi/nm+1$ and use one detection phase ϕr corresponding for instance to the detection of σx and σy. After detecting p qubits in state $j=o$ and N-p in state $j=1$ the inferred photon number distribution has become the distribution maximum is obtained by computing the derivative of $p(n|p;N-p)$ versus n. The derivatives cancels for $X=p/N$ and the photon number n-max satisfies. To estimate the width of the inferred photon number distribution we compute its second derivative at $X=p/N$ and we get the Taylor expansion of the photon number distribution around its maximum.

Fisher information & Cramer-Rao bound

The known probability law for the estimators depends upon an unknown parameter θ of one of the estimators which might also be a vector. For reasons made clear the probability function is called the likelihood of θ of one estimator corresponding to result of the other. The measurement of probability brings an information about θ that we want to quantify. We are going to call estimator θx a function which associates to each x result an estimation of the true θ. The variance of θx averaged over measurements defines the estimator precision. This variance has a lower limit independent of the estimator called the cramer-Rao bound which is in turn related to a function of the likelihood called the Fisher Information. Measuring a random variable X yields a result x.

The known probability law $p(x/\theta$ of X depends upon an unknown parameter θ which can also be a vector. For reasons made clear the function $p(x/\theta$ is called the likelihood of θ corresponding to result x. The measurement of X brings an information about θ that we want to quantify. We are going to call estimator θx a function which associates to each x result an estimation of the true θ. The variance of θx averaged over measurements defines the estimator precision. This variance has a lower limit independent of the estimator called the cramer-Rao bound which is in turn related to a function of the likelihood called the Fisher Information. First we consider unbiased estimators, whose average over a large number of measurements yields the true value of θ. We then use the identity $dp/d\theta = p \, dLogp/d\theta$ which leads to:

$\int \theta(x) - \theta t(\ p \ dLogp/d\theta t(\ dx = 1$

Introduction to superconducting qubits

The physics of superconducting qubits has made tremendous progresses during the last years. Circuits made of Josephson junctions have been turned into artificial atoms which can be manipulated and measured by methods similar to the ones previously developed in ion trap or CQED physics. Here we are going to focus on one aspect of this physics, namely the use of Josephson qubits to prepare and reconstruct non classical fields of radio-frequency resonators/circuit QED and we will compare this physics with CQED.

Ramsey interferometer

Ramsey interferometer with the two cavities R1 et R2 sandwiching the cavity C containing the field to be measured. The atom with two levels g and e qubit in states j=0 and j=1 respectively, prepared in e, is submitted to classic pulses in R1 and R2, the second having ϕr phase difference with the first. The probabilities to detect the atom in g/j=0 and e/j=1 when C is empty are $Pj=\cos^2(\phi r-j\pi(/2$ The Pj probabilities oscillate between 0 and 1 with opposite phases when ϕr is swept.

Probabilities Pe or Pg=1-Pe for finding atom in e or g oscillate versus ϕ. The phase of the atomic fringes and their amplitude depend upon the state of field in C, which affect in different ways the probability amplitudes associated to states e and g. If C is non-resonant with the atomic transition and contains n photons, the atomic dipole undergoes in C a phase shift θ. The fringes are shifted and the Pj probabilities become

$Pj=\cos^2(\phi r+\theta-j\pi(/2$. Detecting the Ramsey signal with phase ϕr amounts to choosing a detection direction of the qubit Bloch vector in the equatorial plane of the Bloch sphere. The phase-shift per photon is set to distinguish photon numbers from 0 to 7, each one corresponding to a different direction of the Bloch vector. Ramsey pulse rotates Bloch vector by $\pi/2$ around vx. Qubit phase shift ϕc during C crossing amounts to Bloch vector rotation around vz. Second Ramsey pulse exercise $\pi/2$ rotation around vu whose direction depends on R2-R1 relative phase ϕr.

Useful formula

For a Pauli operator $\exp(-i\phi\delta u=\cos\phi$ lc- i $\sin\phi\delta u$; hence the rotation induced by Ramsey interferometer would be $R=\exp(-i\pi/4$ $\delta u($ $\exp(-i\phi/2$ $\delta z($ $\exp(-i\pi/4$ δx; The part two term two of the equation is the cavity phase shift. Coupling a qubit to a quantized field mode: The Jaynes-Cummings Hamiltonian Matter/ Classic current Radiation/ Quantum field mode Semi-classic Quantum/qubit Quantum field mode Coherent states Matter / Quantum/qubit Radiation/ Classic field Qubit rotations full quantum Quantum/qubit Quantum field mode Cavity QED

Josephson junction

A Josephson junction is known with two pieces of superconducting metal separated by a few nanometer wide insulating barrier through which Cooper pairs tunnel. When the system isolated, we can define the number of Cooper pairs and the quantum phases of the Cooper pair wave functions on the two sides of the JJ. The charge and phase differences are the essential parameters describing the properties of the JJ. B.Josephson has shown in 1962 that a current $IJ=Io\ sin\sigma$ with σ defined as quantum phase of the Cooper pair wave functions.

Quntum optics in cavity

One of the most subtle problems in the physics of this century is the relation between the macroscopic world, described by classical physics, and the microscopic world, ruled by the laws of quantum physics. Among the several questions involved in the quantum-classical transition, one stands out in a striking way.

As pointed out by Einstein in a letter to Max Born in 1954 ,it concerns the in-existence at the classical level of the majority of states allowed by quantum mechanics, namely coherent superpositions of classical distinct states. Indeed, while in the quantum world one frequently comes across coherent superpositions of states like in Young's two-slit interference experiment, in which each photon is considered to be in a coherent superposition of two wave-packets ,centered around the classical paths which stem out of each slit, one does not see macroscopic objects in coherent superpositions of two distinct classical states ,like a stone which could be at two places at the same time.

There is an important difference between a state of this kind and one which would involve just a classical alternative: the existence of quantum coherence between the two local states would allow in principle the realization of an interference experiment, complementary to the simple observation of the position of the stone. We know all this already from Young's experiment: the observation of the photon path that is, a measurement which is able to distinguish through which slit the photon has passed unavoidably destroys the interference fringes. If one assumes that the usual rules of quantum dynamics are valid up to the macroscopic level, then the existence of quantum interference at the microscopic level necessarily implies that the same phenomenon should occur between distinguishable macroscopic states. This was emphasized by Schrödinger in his famous cat paradox. An important role is played by this fact also in quantum measurement theory, as pointed out by Von Neumann. Indeed, let us assume for instance that a microscopic two-level system (states $|+$ and $|-$) interacts with a macroscopic measuring apparatus,in such a way that the pointer of the apparatus points to a different and classical distinguishable position for each of the two states, that is, the interaction transforms the joint atom-apparatus initial state into $|+> |+>' |\uparrow >, |-> |->' |\uparrow >$. The linearity of quantum mechanics implies that, if the quantum system is prepared in a coherent superposition of the two states, say $|\psi\} = |+\}+|-\}/\sqrt{2}$, the final state of the complete system should be a coherent superposition of two correlated states, each of which corresponding to a different position of the pointer:

$$(1/\sqrt{2})(|+>+|->) |\uparrow > (1/\sqrt{2})(|+>' |\uparrow >+|->' |\uparrow >) = (1/\sqrt{2})(|\uparrow >' + |\uparrow >') (1,$$

where in the last step it was assumed that the two-level system is incorporated into the measurement apparatus after their interaction for instance, an atom which gets stuck to the detector. One gets, therefore, as a result of the interaction between the microscopic and the macroscopic system, a coherent superposition of two classical distinct states of the macroscopic apparatus. This is indeed the situation in Schrödinger's cat paradox: the cat can be viewed as a measuring apparatus of the state of a decaying atom, the state of life or death of the cat being equivalent to the two positions of the pointer. This would imply that one should be able in principle to get interference between the two states of the pointer:

it is precisely the lack of evidence of such phenomena in the macroscopic world which motivated Einstein's concern. Faced with this problem, Von Neumann introduced through his collapse postulate two distinct types of evolution in quantum mechanics: the deterministic and unitary evolution associated to the Schrödinger equation, which describes the establishment of a correlation between states of the microscopic system being measured and distinguishable classical states for instance, distinct positions of a pointer of the macroscopic measurement apparatus; and the probabilistic and irreversible process associated with measurement, which transforms the correlated state into a statistical mixture.

This separation of the whole process into two steps has been the object of much debate; indeed, it would not only imply an intrinsic limitation of quantum mechanics to deal with classical objects, but it would also pose the problem of drawing the line between the microscopic and the macroscopic world. Several possibilities have been explored as solutions to this paradox, including the proposal that a small non-linear term in the Schrödinger equation, although unnoticeable for microscopic phenomena, could eliminate the coherence between macroscopic states, thus transforming the quantum superpositions into statistical mixtures. The non-observability of the coherence between the two positions of the pointer has been attributed both to the lack of non-local observables with matrix elements between the two corresponding states as well as to the fast decoherence due to dissipation. This last approach has been emphasized in recent years: decoherence follows from the irreversible coupling of the observed system to a reservoir. In this process, the quantum superposition is turned into a statistical mixture, for which all the information on the system can be described in classical terms, so our usual perception of the world is recovered. Furthermore, for macroscopic superpositions quantum coherence decays much faster than the physical observables of the system, its decay time being given by the dissipation time divided by a dimensionless number measuring the separation between the two parts. The statement that these two parts are separated in macroscopic terms implies that this separation is an extremely large number. Such is the case for biological systems like cats made of huge number of molecules. In the simple case mentioned by Einstein,of a particle split into two spatial separated wave packets by a distanced, the dimensionless measure of the separation is $[d/\lambda dB[^2$, where λdB is the particle de Broglie wavelength. For a particle with mass equal to1g at a temperature of 300K, and d=1 cm, this number is about $10^{\wedge}40$, and the decoherence is for all purposes instantaneous. This would provide an answer to Einstein's concern: decoherence of macroscopic states would be too fast to be observed. The study of the interaction between atoms and electromagnetic fields in cavities can help us understand some aspects of this problem. In fact, many recent contributions in the field of quantum optics have led not only to the investigation of the subtle frontier between the quantum and the classical world, but also of hitherto unsuspected quantum mechanical procedures like teleportation. Research on quantum optics is therefore intimately entangled with fundamental problems of quantum mechanics. The whole area of cavity quantum electrodynamics is a very recent one. It concerns the interactions between atoms and discrete modes of the electromagnetic field in a cavity, under conditions such that losses due to dissipation and atomic spontaneous emission are very small. Usually, one deals with atomic beams crossing cavities with a high quality factor Q defined as the product of the angular frequency of the mode and its lifetime, $Q=\omega\tau$.

The atoms, prepared in special states and detected after interacting with the field, serve two purposes: they are used to manipulate the field in the cavity, so as to produce the desired states, and also to measure the field. Several factors contributed to the development of this area. The production of superconducting Niobium cavities, with extremely high quality factors, up to the order of 10^{10}, allows one to keep a photon in the cavity for a time of the order of one second. New techniques of atomic excitation alkaline atoms, like Rubidium and Cesium, are frequently used for this purpose to highly excited levels principal quantum numbers of the order of 50, and with maximum angular momentum $l=n-1/$ the so-called planetary Rydberg atoms – have led to the production of atomic beams that interact strongly even with very weak fields, of the order of one photon, due to the large magnitude of the relevant electric dipoles. Besides,the lifetime of these states is large–of the order of the millisecond–which might be understood semi-classical, from the correspondence principle which should be in place for n 50): the electron is always very far away from the nucleus, and therefore its acceleration would be small,implying weak radiation and a long lifetime. Looking into the problem of the classical limit of quantum mechanics will indeed provide us with a useful thread, a leitmotif which will lead us to many important techniques of quantum optics.

Coherent states

We have seen that the average value of the electric field operator vanishes in a Fock state. Therefore, we cannot associate Fock states to classical fields with amplitude different from zero, and well-defined phase. We should look for quasi-classical field states. Provided that the average value of the electromagnetic field in these states, which we denote by $|\alpha>$, coincides with the classical expression for an electromagnetic field with complex amplitude α: $<\alpha|E(r)|\alpha>=\sqrt{\omega/V}[u(r)\varepsilon\alpha$

+ c.c.

The states |n> are the so-called Fock states, and have a well-defined number of photons. The state corresponding to n=0 is the vacuum state. It is easy to show from the above relations that $|n=a'n/\sqrt{n!}|0>$. The electric field is expressed in terms of the cancel out and create operators by $E(r)=E\omega[au(r)\varepsilon +a'u(r)]$ where u(r) is a function which describes the spatial dependence of the field mode, is the polarization vector, and $E_w= \omega/V$ is the field per photon. Here $V=\int|u(r)|^2 d^3r$, is the effective volume of the mode, defined so that the expectation value of the electromagnetic energy in the vacuum state, $1/4\pi \int<0| [E(r)]^2 |0> d^3r$, is equal to the zero→ point energy $\omega/2$.

A special role will be played in the following by the phase displacement operator:

$U(\theta)=\exp(-i\theta N)$

The expectation value of the electromagnetic energy in the state |α> coincides with the classical expression for this energy, expressed in terms of the complex amplitude α, at least in the limit when $|\alpha|>>1$. It is easy to show that the classical energy, when the electric field expressed in terms of α, would be

$E_c= \omega |\alpha|^2$

$<\alpha|a|\alpha>=\alpha$

Provided that the expectation value of the electromagnetic energy in the state |α> coincides with the classical expression for this energy, expressed in terms of the complex amplitude α, at least in the limit when $1<< \alpha$.

Luiz Davidovich

Quantum Mechanics

Quantum Phase

A quantum state can be expressed as a linear combination of components of other eigenstates; every component eigenstate has an associated phase. It is this phase which gives the wavefunction its wavelike character. In order for the components to combine together correctly to produce a superposition state, they should be in the same phase/coherent. Were the phases of the states be altered, the coherent phase relationships between the components would be destroyed.

Quantum Mechanics

In western view of quantum mechanics focus is too much on analysis and tends to lead to fragmentation, what is meant by fragmentation is not just division or distinction, because parts and the whole are correlative concepts, part is a part because it is part of a whole like a machine or a watch, the fragment is something you see when you break up the whole, so if you smash the watch you get the fragments, the western worldview aims at getting the true parts of the universe but in some ways perhaps it gets the fragments, to some extent. So if you break it up into fragments then you have it confused, you are going to treat that as if it is separate or not, and also you are going to unite what's in fragment, when it's not united, so it leads to confusion. you get confused about the part and the whole, because you take the fragment as an independent whole. The observer is the intrinsic part of the whole, that's what quantum mechanics teaching us, that the observing instrument is just as much part of the whole, and therefore because of the possibilities of these non-local interactions in quantum mechanics when an observation is made the two systems are not really distinct, therefore they participate in each other and therefore you can not get an ambiguous meaning to the measurement.

The same happens between human beings, If somebody tries to measure somebody else trying talk to him there is a mutual change which make it impossible to get unambiguous attribution of quality; we are participating together, that's what happens in quantum mechanical observation. In 1952 I developed an interpretation, I said that the electron is a particle for example, then it has a quantum field represented by its wave function and this field and the particle are together, so it is a new kind of field, We know in classic view we have different kind of field like electric, magnetic, for example, the electromagnetic fields makes radio waves radiating through space, but the quantum field is different, it has some similarities but it's different because the effect of quantum field depends only on the form and not on the intensity. The quantum field will be capable of sometimes of spreading out and the electron at far away could move with the same energy as if it were close, this would be a kind of explanation of this discrete quantum process. So I have a field that its effect doesn't drop off, The field might drop off but its effect does not, the effect depends only on the form, not on the intensity, it is very common but we often don't pay attention to it, If you take for instance a radio wave, its effect falls up, now imagine a ship guided by a radar or automatic pilot the guidance doesn't depend on the intensity of the wave, it depends on the form which we may say carries information, the word information has the two words, "To put form In". So it is like if you have a TV set and you go far away from the antenna or the place where they put out the broadcast, it doesn't mean that you don't get the broadcast, you just need a receiver; a Sensitive receiver. As long as it receives it is the same program, what happens is that the form of radio wave puts form into the current flowing into the receiver, the energy comes from the receiver not from the radio wave. The radio wave is not pushing the ship around in a mechanical sense, the ship is moving on its own term and responding with the form, so the radio wave is guided and form giving shape to its motion. The form of the DNA determines its character RNA and determines the making of proteins, In human experience we observe the same phenomena : people don't push each other around except if they are violent, they depend on the form to communicate and people move around because of that, So this is the most common form of experience and the mechanical business of pushing and pulling here is more limited. Perhaps this form is fundamental and the electron respond with this form, this not only explains the interference but also why electrons act like waves, and explains the non-local business and so on, and explains the superconductivity as the electrons moving by the common pool of information just as the ballet dancers do. That means we have quite a different principle of explanation because this wave function which operates through form is closer to life and mind, the basic quality of mind is that it responds to form and not to the substance. One might think from what has been said that the physical universe is more about information rather than substance, One might say both, but information contributes in a fundamental way to the quality of the substance.

We can discuss the mathematics given the probability that certain results would be obtained, this gives an intelligible explanation; requires you to accept new principles and you would have to say this wave field will perhaps have a more substantial basis which we don't know that would carry in. You might regard it as a kind of speculation which is not tied to experimental facts perhaps, but it is important to make it intelligible, also to show the connection between this and the whole range of experiences. This view may or may not give a same experimental predictions as the classical view, but the experimental prediction is only one of the functions of the theory, it enables you to understand whats going on to make it intelligible, When the general audience is presented with this view toward the universe which is often based on the interpretation of the quantum mechanics, they might acknowledge that this is sort of a minority kind of interpretation, that's because other interpretations that are known to them are not intelligible, it's so abstract and difficult that they cant really understand, however this interpretation makes everything more accessible to more people and perhaps shows the connection of different fields. The act of information is quite far into classical physics, the thing makes classical physics is not just the form of newtons law, but what you say about the forces, if you say that they are this character of information it changes, but the entirely non-classical concept which is the activity of information contributes to the property of substance in its foundation. Our concern should be whether we take the wave function as the whole description or not; I have this particle, and the wave function for it would be as the field of information that acts on the particle, the wave function which is a mathematical representation of the field of information, in the case of one particle it's like a wave but it's a wave that acts according to its form and not according to its intensity. We might regard the wave function as a part of reality, and this might be the case, we could make an analogy to society, one would say that the society consist of people and interrelated but another point of view might say they are interrelated but through information exchange, that's crucial, without that society collapse. If you compare it to society you can have every individual follow his own pool of information or can have people trying to move together within a common pool, I think it's essential to have coherence and harmony that the whole society could move together with this common pool of information which is established by exchange and dialogue.

The general trend hasn't got very far toward that direction because everything has been divided into nations and religions and other kinds of groups which behave as if they are independent where they are not, people have to give all that up and they might find it hard, to deal with the ecological problem I think people have to give the great deal of that up. One might think I am moving my emphasis from persons, individuals or the divided parts to the information flow or the information field of society; that might be right in some sense but I say At the same time each individual contains the whole information field of society in his own way; it's in his mind, it's in his brain, Each individual in him or her has the whole human experience or knowledge; What I'm trying to say is that individuals come from society but individuals together form society, Now we have so many individuals each with his own view, come into clash, they have gotta be able to talk about it, to have dialogue and to entertain each others view to look at it calmly that each one can look at all the views, so each individual holds all the views and holds the whole and he doesn't necessarily agree with them but out of that what he merge with is the common pool of information which guides the society. Quantum mechanics gives the probability of an experimental result, the decay of an atomic nucleus and the fact that it decays at one moment and not another cannot be properly pictured within the theory; however, It can enable you to predict the results of various experiments. Physics has changed from its earlier form, when it tried to explain things and give some physical picture; Now the essence is regarded as mathematical, It's felt the truth is in the formulas, Now they may find an algorithm by which they explain a wider range of experimental results. In the Fifties, when I sent my book around to various physicists-including Bohr, Einstein, and Pauli--Bohr didn't answer, Pauli liked it. Einstein sent me a message that he'd like to talk with me. When we met he said the book had done about as well as you could do with quantum mechanics, however he was not convinced it was satisfactory enough to be considered as a theory, His objection was not that it was statistical, He felt it was a kind of abstraction; quantum mechanics got correct results but left out much that would have made it intelligible. I came up with the causal interpretation, that the electron is a particle, but it also has a field around it. The particle is never separated from that field, and the field affects the movement of the particle in certain ways. Einstein liked it, however, the interpretation had this notion of action at a distance: Things that are far away from each other profoundly affect each other; He believed in local action, I didn't come back to this implicate order until the Sixties, when I got interested in notions of order. I realized then the problem is that coordinates are still the basic order in physics, whereas everything else has changed. When it comes to enfoldment, Everybody has seen an image of You fold up a sheet of paper, turn it into a small packet, make cuts in it, and then unfold it into a pattern, The parts that were close in the cuts unfold to be far, This is like what happens in a hologram. Enfoldment is really very common in our experience. All the light in this room comes in so that the entire room is in effect folded into each part. If your eye looks, the light will be then unfolded by your eye and brain. As you look through a telescope or a camera, the whole universe of space and time is enfolded into each part, and that is unfolded to the eye. With an old-fashioned television set that's not adjusted properly, the image enfolds into the screen and then can be unfolded by adjustment.

To map the coordinates and order tie in with enfoldment; it

doesn't have to be necessarily straight lines. They are a way of mapping space and time, Since space-time may be curved, the lines may be curved as well. It became clear that each general notion of the world contains within it a specific idea of order. The ancient Greeks had the idea of an increasing perfection from the earth to the heavens. Modern physics contains the idea of successive positions of bodies of matter and the constraints of forces that act on these bodies. The order of perfection investigated by the ancient Greeks is now considered irrelevant. The most radical change in the notion of order since Isaac Newton came with quantum mechanics. The quantum-mechanical idea of order contradicts coordinate order because Heisenberg's uncertainty made a detailed ordering of space and time unlikely. When you apply quantum theory to general relativity, at very short distances like ten to the minus thirty-three centimeters, the notion of the order of space and time breaks down. To replace that with some other sense of order First you have to ask what we mean by order, Everybody has some tacit notion of it, but order itself is impossible to define, Yet it can be illustrated; In a photograph any part of an object is imaged into a point, This point-to-point correspondence emphasizes the notion of point as fundamental in sense of order, Cameras now photograph things too big or too small, too fast or too slow to be seen by the naked eye. So our image is the lens, the apparatus suggesting the point, The point in turn suggests electrons and particles And the track of particles on the photograph. Now what instrument would illustrate wholeness? Perhaps the holograph; Waves from the whole object come into each part of the hologram; This makes the hologram a kind of knowledge of the whole object. If you examine it with a very narrow beam of laser light, it's as if you were looking through a window the size of that laser beam. If you expand the beam, it's as though you are looking through a broader window that sees the object more precisely and from more angles, But you are always getting information about the whole object, no matter how much or little of it you take. But let's put aside the hologram because that's only a static record, Returning to the actual situation, we have a constant dynamic pattern of waves coming off an object and interfering with the original wave, Within that pattern of movement, many objects are enfolded in each region of space and time. Classical physics says that reality is indeed little particles that separate the world into its independent elements, Now I'm proposing the reverse, that the fundamental reality is the enfoldment and unfoldment, and these particles are abstractions from that. We could picture the electron not as a particle that exists continuously but as something coming in and going out and then coming in again. If these various condensations are close together, they approximate a track; The electron itself can never be separated from the whole of space, which is its ground. About the time I was looking into these questions, a BBC science program showed a device that illustrates these things very well, It consists of two concentric glass cylinders, Between them is a viscous fluid, such as glycerin, If a drop of insoluble ink is placed in the glycerin and the outer cylinder is turned slowly, the drop of dye will be drawn out into a thread,

Eventually the thread gets so diffused it cannot be seen, At that moment there seems to be no order present at all; Yet if you slowly turn the cylinder backward, the glycerin draws back into its original form, and suddenly the ink drop is visible again, The ink had been enfolded into the glycerin, and it was unfolded again by the reverse turning. Suppose you put a drop of dye in the cylinder and turn it a few times, then put another drop in the same place and turn it. When you turn the cylinder back, wouldn't you get a kind of oscillation? Yes, you would get a movement in and out, We could put in one drop of dye and turn it and then put in another drop of dye at a slightly different place, and so on; The first and second droplets are folded a different number of times, If we keep this up and then turn the cylinder backward, the drops continually appear and disappear, So it would look as if a particle were crossing the space, but in fact it's always the whole system that's involved. We can discuss the movement of all matter in terms of this folding and unfolding, which I call the holo-movement which may lie outside of time as we ordinarily know it. If the universe began with the Big Bang and there are black holes, then we must reach places where the notion of time and space breaks down. Within the singularity none of the laws as we know them apply, There are no particles; they are all disintegrated, There is no space and no time, Whatever is, would be beyond any concept we have at present. The present physics implies that the total conceptual basis of physics must be regarded as completely inadequate. The grand unification of the four forces of the universe could be nothing but an abstraction in the face of some further unknown. I propose something like this: Imagine an infinite sea of energy filling empty space, with waves moving around in there, occasionally coming together and producing an intense pulse. Let's say one particular pulse comes together and expands, creating our universe of space-time and matter, But there could well be other such pulses; To us, that pulse looks like a big bang; In a greater context, it's a little ripple. Everything emerges by unfoldment from the holo-movement, then enfolds back into the implicate order; I call the enfolding process "implicating," and the unfolding "explicating." The implicate and explicate together are a flowing, undivided wholeness, Every part of the universe is related to every other part but in different degrees. There are two experiences: One is movement in relation to other things; the other is the sense of flow The movement of meaning is the sense of flow, even in moving through space, there is a movement of meaning. In a moving picture, with twenty-four frames per second, one frame follows another, moving from the eye through the optic nerve, into the brain. The experience of several frames together gives you the sense of flow; This is a direct experience of the implicate order. In classical mechanics, movement or velocity is defined as the relation between the position now and the position a short time ago, What was a short time ago is gone, so you relate what is to what is not, This isn't a logical concept; In the implicate order you are relating different frames that are co-present in consciousness, A moment contains flow or movement, The moment may be long or short, as measured in time; So a moment enfolds all the past, but the recent past is enfolded more strongly, At any given moment we feel the presence of all the past and also the anticipated future, It's all present and active. I could use the example of the cylinder again; Let's say we enfold one droplet h times, Then we put another droplet in and enfold it N times, The relationship between the droplets remains the same no matter how thoroughly they are enfolded, So as you unfold, you will get back the original relationship. Imagine if we take four or five droplets--all highly enfolded--the relationship between them is still there in a very subtle way, even though it is not in space and not in time. But, of course, it can be transformed into space and time by turning the cylinder. The best metaphor might involve memory; We remember a great many events, which are all present together, Their succession is in that momentary memory, We don't have to run through them all to reproduce that time succession since we already have the succession. Much of our experience suggests that the implicate order is natural for understanding; When you are talking to somebody, your whole intention to speak enfolds a large number of words. You don't choose them one by one. There are any number of examples of the implicate order in our experience of understanding, Any one word has behind it a whole range of meaning enfolded in thought. Understanding is unfolded in each individual, Clearly, it's shared between people as they look at one object and verify that it's the same, So any high level of that is a social process. There might be some level of sensorimotor perception that is purely individual, but any abstract level depends on language, which is social. The word, which is outside, evokes the meaning, which is inside each person, Meaning is the bridge between understanding and substance. Any given array of matter has for any particular mind a significance. The other side of this is the relationship in which meaning is immediately effective in matter. Suppose you see a shadow on a dark night; If it means "assailant," your adrenaline flows, your heart beats faster, blood pressure rises, and muscles tense, The body and all your thoughts are affected, everything about you has changed. If you see that it's only a shadow, there's an abrupt change again. That is an example of the implicate order: Meaning enfolds the whole world into me, and vice versa- that enfolded meaning is unfolded as action, through my body and then through the world. Physicists are very skeptical that the implicate order is worth investigating, The most convincing thing would be to develop the theory in mathematical terms and make some experimental predictions. A few years ago The New York Times noted that some physicists were critical of grand unification theory, saying that not much had been achieved, Defenders of grand unification theories said it would take about five years to see results, It seems that people are ready to wait four or five years for results if you've got formulas. If there are no formula, no mathematical representation, they don't want to consider it. Formulas are means of talking utter nonsense until you understand what they mean. Every page of formulas usually contains six or seven arbitrary assumptions that take weeks of hard study to penetrate. Younger physicists usually appreciate the implicate order because it makes quantum mechanics easier to grasp. By the time they're through graduate school, they've become dubious about it because they've heard that hidden variables are of no use because they've been refuted. At this point, I think that the major issue is mathematics. In super symmetry theory an interesting piece of mathematics will attract attention, even without any experimental confirmation. David Bohm

The Density Probability of the Schrödinger equation

Quantum theory consists of two important parts:
1 The Density Probability of the Schrödinger Equation
2 The Von Neumann Measurement Postulate

The Density Probability of the Schrödinger Equation

Spatial motion of free quantum particles is described by the Schrödinger wave equation. Typical solutions of this equation show an unlimited growth of the wave packet width. If the particle is illuminated then the scattered light will show the trajectory of the quantum particle. If a single photon of incoming wave number has scattered with the final wave number then the scatterer's wave function obtains a unitary factor. where the scattering angle θ between incoming wave number and the final wave number, resp., is distributed according to the modulus square of the scattering amplitude. These processes interrupt the otherwise deterministic evolution of the wave function. We shall assume a certain diffuse light of different wave number. Then the scattering processes will occur randomly at a given rate. To get an easy insight into the resulting stochastic process, we shall assume that $\lambda = 2\pi/k$ is much bigger than the wave packet width and, furthermore, that the repetition frequency of scatterings is big as compared to the time scale of the free dynamics of the particle. What follows will describe the dynamics of the illuminated particle. The particle's motion is influenced by the random force which is a certain stationary white noise. Hence, the linear stochastic might not be suitable to represent the experienced trajectory of the quantum particle. The jumps of the wave function assumes a particular classification of the photon states. Indeed, the final states as well as the initial ones have been classified according to their momenta. Obviously, trajectories can not be observed via the scattering angles of the photons. They are only observed by identifying the position where the scattered light has emerged. One might think of a lense inserted on the path of each scattered photon, making an optical map of the scatterer particle. Here we introduce the mathematical model of the above set up by the special Fourier transform of the scattering amplitude to present the influence of a single scattering process on the particle's wave function. This jump differs very much from the previous one in a sense that is nonlinear.;In the same approximation that we assumed for the previous process, this nonlinear one leads to the following counterpart of a henomenological equation. To make a formal comparison of this last procedure to the ordinary Schrödinger, The main difference would be the lack of the imaginary factor in the second term as if the random force had become pure imaginary. The third term then comes in just to restore normalization of the wave function.

Lajos Diosi

The evolution of the wave function described by the Schrödinger equation has two key features

1 It is a reversible process, so one can always undo the process in order to return to the initial state from the final state

2 It is a deterministic evolution in which the final state is

uniquely determined by the initial state if the Hamiltonian is known.

The Schrödinger equation
ih dω/dt = - h²/2m d²ω/dx²
The wave function when rotated $\pi/2$ in the complex plane, changes in time in proportion to its curvature in space and in inverse proportion to its mass.

The time independent Schrödinger equation
For one dimension is of the form -h²/2m d²ωx/dx²+Ux ωx; where Ux is the potential energy. The time dependent Schrödinger equation For one spatial dimension is of the form (-h²/2m)∂²ω(x,t)/∂x²+Ux ω(x,t) For a free particle where U(x) =0 the wave function solution can be put in the form of a plane wave ω(x,t)=Aeikx. For other problems, the potential Ux serves to set boundary conditions on the spatial part of the wavefunction and it is helpful to divide the equation into the time-independent Schrödinger equation and the relationship for time evolution of the wave function.

Time Evolution: Hω=ih dω/dt
Time Independent Equation -h ²/2m ∂²ωx/∂x²+Ux ωx

The Phase Factor
The solution of the Schrödinger equation yields a wave function whose phase is immaterial when it comes to measurement. The phase factor α, is space-time independent, to α(r,t). If the

86

original Schrödinger equation contains the potential energy qφ, the phase factor needs to undergo the transformation α(r,t) which is space-time dependent in order to cancel the term r,t. If the phase factor slowly changes its direction such that it traces out a path that returns it to its original orientation; the Hamiltonian varies accordingly. The eigenstates, remaining referenced to the phase factor throughout its slow evolution. Upon completion of the path, the phase factor has returned to its original orientation, but has the eigenstate also recovered its original form, or has it acquired a spin due to its adiabatic passage around the path.

Free Particle Wave Equation

For a free particle the time-dependent Schrödinger equation takes the form:

$-h^2/2m\ \partial^2\omega(x,t)/\partial x^2 = ih\ \partial\omega(x,t)/\partial t$; Presuming that the wave function represents a state of definite energy E, the equation can be divided by the requirement:

$E\omega = -h^2/2m\ \partial^2\omega(x,t)/\partial x^2$ $E\omega = i\hbar\ \partial\omega(x,t)/\partial t$

Arriving at Time Dependent Schrödinger

The time-dependent Schrödinger equation is a partial differential equation, a complete understanding of which requires more mathematical preparation than we are assuming here. Fortunately, the majority of interesting problems in quantum theory do not require use of the equation in its full. By far the most interesting states of any quantum system are those states in which the system has a definite total energy, and it turns out that for these states the wave function is a standing wave, analogous to the familiar standing waves on a string. When the time-dependent Schrödinger equation is applied to these standing waves, it reduces to a simpler equation called the time-independent Schrödinger equation. We will need this time-independent equation, which will let us find the wave functions of the standing waves and the corresponding allowed energies. Because we will be using only the time-independent Schrödinger equation we will often refer to it as just the Schrödinger equation. The wave function of a particle of fixed energy E could be written as a linear combination of wave function s of the form $\omega(x,t)$-Aei(kx-wt) (1

representing a wave travel in the positive x direction, and a corresponding wave travel in the opposite direction, so giving rise to a standing wave, this being necessary in order to satisfy the boundary conditions. The wave function above considered as being the appropriate wave function for a free particle of momentum p-hk and energy E- hw. We can note that $\partial^2\omega/\partial x^2 = -k^2\omega$ (2

can be written, using E-p^2/2m- h^2k^2/2m: $- h^2$/2m $\partial^2\omega/\partial x^2$- p^2/2m ω (3

using E-hw: $ih\partial\omega/\partial t$- hw-E$\omega$ (4

Where there is both a kinetic energy and a potential energy present, then E-p^2/2m+V(x) so that

Eω-p^2/2m ω +V(x)ω (5

with ω is now the wave function of a particle moving in the presence of a potential V(x). But if we assume that the results

Eq.(3) and Eq.(4) still apply in this case then we have

$$-h^2/2m\ \partial^2\omega/\partial x + V(x)\omega - i\ \partial\omega/\partial t \qquad (6$$

Even though this equation does not look like the familiar wave equation that describes, for instance, waves on a stretched string, it is nevertheless referred to as a wave equation as it can have solutions that represent waves propagating through space. In general, the solutions to the time dependent Schrödinger equation will describe the dynamical behaviour of a particle, in some way; solving Schrödinger equation, what we get is a wave function $\omega(x,t)$ that tells us how the probability of finding a particle in some region in space varies as a function of time.

The Time Independent Schrödinger Equation

The time dependence enteres into the wave function via a complex exponential factor exp[$-$ h iEt/ This suggests that to extract this time dependence we guess a solution to the Schrödinger wave equation of the form $\omega(x,t)-\omega(x)e - ihEt/$ (1

i.e. where the space and the time dependence of the complete wave function are contained in separate factors. If we substitute this trial solution into the Schrödinger wave equation, and make use of the meaning of partial derivatives; The quantity E; the energy of the particle, is a free parameter in this equation. To determine the wave function for a particle with some specific value of E that is moving in the presence of a potential V(x), what we do is to substitute the value of E into the equation with the appropriate V(x), and solve for the corresponding wave function. Writing ωE(x) as the solution accompanied with a particular value of E, It turns out that it is not quite as simple as we supposed that to be. The wave function ωE(x) plus its derivative should be continuous. The probability interpretation of the

wave function along with its continuity leads to the quantization of energy.

The relation between velocity and momentum in non-relativistic quantum mechanics In quantum mechanics, the velocity v, like the position q and the momentum p, is an operator. It is defined by the relation:

$v = i/h$ (H,q   (1

where H is the hamiltonian operator, and [a,b] =ab − ba is the commutator of the operators a and b. In non-relativistic quantum mechanics,

$H = -h^2/2m \, \nabla^2 q + V(q$ (2

corresponding to the time dependent Schrödinger equation

$ih\partial \omega dt = H\omega$ (3

Hence, substituting this expression for H in Eq.1, one finds that the velocity operator is given by: $v = p/m$ (4 where $p = -ih\nabla$ q (5 is the momentum operator.

The Hamiltonian Operator

$H = -h^2/2m \, d^2\omega/dx^2 + Vx$

Operators In Quantum Mechanics

With each measurable parameter in a physical system is a quantum mechanical operator. Such operators arise because in quantum mechanics you are describing nature with waves (the wave function) rather than with discrete particles whose motion and dynamics can be described with the deterministic equations of Newtonian physics. Part of the development of quantum mechanics is the establishment of the operators associated with the parameters needed to describe the system. It is part of the basic structure of quantum mechanics that functions of position are unchanged in the Schrödinger equation, while momenta take the form of spatial derivatives. The Hamiltonian operator contains both time and space derivatives.

Energy Eigenvalues

To obtain specific values for energy, you operate on the wave function with the quantum mechanical operator associated with energy, which is called the Hamiltonian. The operation of the Hamiltonian on the wave function is the Schrödinger equation. Solutions exist for the time-independent Schrödinger equation only for certain values of energy, and these values are called "eigenvalues" of energy. While the energy eigenvalues may be discrete for small values of energy, they usually become continuous at high enough energies because the system can no longer exist as a bound state. For a more realistic harmonic oscillator potential (perhaps representing a diatomic molecule), the energy eigenvalues get closer and closer together as it approaches the dissociation energy. The energy levels after dissociation can take the continuous values associated with free particles.

The Eigenfunctions

Corresponding to each eigenvalue is an eigenfunction; The solution to the Schrödinger equation for a given energy involves also finding the specific function which describes that energy state. The eigenvalue concept is not limited to energy. When applied to a general operator Q, it can take the form $Q_{op} \omega_i = q_i \omega_i$; If the function ω_i is an eigenfunction for that operator. The eigenvalues q_i may be discrete, and in such cases we can say that the physical variable is "quantized" and that the index i plays the role of a "quantum number" which characterizes that state.

Time Evolution

If y is the wavefunction for a physical system at an initial time and the system is free of external interactions, then the evolution in time of the wave function is given by: $H\omega = ih\, \partial\omega/\partial t$ where H is the Hamiltonian operator formed from the classical Hamiltonian by substituting for the classical observables their corresponding quantum mechanical operators. The role of the Hamiltonian in both space and time is contained in the Schrödinger equation.

Probability Density Function

Most often, the equation used to describe a continuous probability distribution is called a probability density function . Sometimes, it is referred to as a density function. Since the continuous random variable is defined over a continuous range of values; called the domain of the variable, the graph of the density function will also be continuous over that range. The area bounded by the curve of the density function and the x-axis is equal to 1, when computed over the domain of the variable. The probability that a random variable assumes a value between a and b is equal to the area under the density function bounded by a and b.

Probability Density

Equation that predicts both the allowed energies of a system as well as the probability of finding a particle in a given region of space does not yield the probability directly, but rather the probability amplitude which is the probability that a single quantum particle moving in one spatial dimension will be found in a region x[a,b, if a measurement of its location is performed is $P[x(a,b=\int \omega x^2 dx$ which is known as probability density px; the probability that a quantum particle will be found in a very small region dx about the point x is px $dx=\omega x^2 dx$ Since particles can exhibit wave-like behavior, the amplitude or wave function ωx should have a wave-like form. The Schrödinger equation cannot be derived from any more fundamental principle. Recall the de Broglie hypothesis stating that the particle has a wavelength λ given by $\lambda= h/p$ or $p=h/\lambda$, If the particle is a free particle, its potential energy $V(x)=0$, so that its energy is purely kinetic $E=p^2/2m= h^2/2m\lambda^2$; If the amplitude ωx describes a wave, then it should take the mathematical form $\omega x=B\sin 2\pi x/\lambda$,we are considering a wave that is not changing in time here; Consider the cosine form (the same will hold for a sine for as well) and consider the first two derivatives of ωx : $d\omega/dx = -2\pi/\lambda$ $B\sin 2\pi x/\lambda$ $d^2\omega/dx^2 = -4\pi^2/\lambda^2$ $B\cos 2\pi x/\lambda$; therefore ωx and $d^2\omega/dx^2$ are related by $d^2\omega/dx^2 = -4\pi^2/\lambda^2 * \omega x$

Probability density

Density operator for a field mode $\rho = |\omega q><\omega q|$ /pure state; $\rho = \sum \rho i |\omega(iq><\omega(iq$ /mixed state.

Spin Operators and the Pauli Matrices For a spin-1/2 system

The 1/2 refers to the quantum numbers. Spin should be considered as spin angular momentum and is related to the intrinsic magnetic moment of a particle. Unlike orbital angular momentum (l= 0, 1,2 ,....), spin angular momentum can take on integer and half-integer values (s= 0, 1/2, 1, 3/2, 2,...). Half-integer particles are called fermions and integer spin particles are called bosons. The spin operators, Sx,Sy, and Sz represent the effect of a measurement of the spin in one of those three directions, while S^2 describes the measurement of the spin squared of a particle. It turns out that these four operators can be written in terms of Pauli 2 x 2 matrices.

Eigenstates of Sz

Since the z component of a spin-1/2 particle can take on just two values, +h/2 or − h/2 (quantum numbers ms= +1/2 or − 1/2), we often call these states spin up or down but indeed we are referring just to the z component of spin. Since there are two values, we can write these states as two-component column vectors which are eigenstates of Sz:

$|1/2,1/2> = |z+> = |\uparrow>$ $|1/2,-1/2> = |z-> = |\downarrow>$; These states are clearly eigenstates of Sz which can be verified by direct calculation. These states can be explicitly determined by using linear algebra to determine the eigenvalues and eigenvectors of the Sz matrix.

Eigenstates of Sx and Sy

It is also of interest to write down the eigenstates of the other spin operators. We can write such eigenstates in terms of eigenstates of Sz if we like. This is called a choice of basis or choice of basis vectors. In particular, we find that the eigenstates of Sx are: or that We also have that and at last The form of these states should convince you that eigenstates of one component of spin will not be eigenstates of the other components of spin.

Classic Standing Waves

By superposing two traveling waves, we form a standing wave. If we imagine a string clamped between two fixed points separated by a distance a, we can ask: What are the possible standing waves that can fit on the string? The answer is that a standing wave is likely, provided that it has nodes at the two fixed ends of the string. The distance between two adjacent nodes is 2λ, so the distance between any pair of nodes is an integer multiple of this, $2n\lambda$. Therefore standing wave fits on the string provided $2n\lambda = a$, for some integer n; that is to say; if $\lambda = 2a/n$ where n=1,2,......

The wave function probability density
$\omega(x,t)^2$ is the probability density associated with a quantum wave function. for the complex standing wave this has a remarkable property; $\omega(x,t)^2 = \omega x^2$ e-iwt^2 For a quantum standing wave, the probability density is independent of time. so the equation above simplifies to $\omega(x,t)^2 = \omega x^2$. This is likely because the time dependent part of the wave function; e-iwt=coswt-isinwt, is complex with two parts that oscillate $\pi/2$ out of phase; when one is growing, the other is shrinking in such a way that the sum of their squares is constant. Thus for a quantum standing wave, the distribution of matter is time independent or stationary. For this reason a quantum standing wave is often called a stationary state, who are the modern counterpart of Bohr's stationary orbits and are precisely the states of definite energy. Because their charge distribution is static, atoms in stationary states do not radiate. An important practical consequence of this would be that in most problems the only interesting part of the wave function $\omega(x,t)$ is its spatial part ωx. We see that a large part of quantum mechanics is devoted to finding the possible spatial functions and their corresponding energies.

The Von Neumann Measurement Postulate

Time dependent Von Neumann measurement

Consider the time dependent observable $Q(t)=Z \sum ß p(t)$ composed of GMH's projectors. Observe that coarse graining is equivalent to N subsequent measurements of $Q(t_1),Q(t_2).....Q(t_n)$ to perform a la Von Neumann. The time set h is given by the sequence $(ß_1,ß_2.....ß_N)$ of the individual measure outcomes. The time set-dependent state turns out to be the resulting state after the N measures and, furthermore, the probability of the corresponding history can be recognized as the standard probability of the N subsequent wave function collapse. We owe to translate the crucial decoherence criteria as well. Let us consider the most simple case by taking N=2, and consider decoherence between histories $h=(ß_1,ß_2)$ and $h'=(\alpha_1,\alpha_2)$ respectively. Invoking the decoherence functional must vanish for nonidentical time sets. It is now straightforward to prove the following statement: "If the equation for nonidentical time sets is satisfied then the expectation value of $Q(t_2)$ in a Von Neumann measurement at t_2 will be independent of whether $Q(t_1)$ was earlier at $t_1<t_2$ measured or was not measured".

Group-Measure space von Neumann

Given a measure preserving action Γ X of an infinite group, one can construct an associated group-measure space von Neumann algebra, the crossed product M=L(X) Γ.

If Γ is infnite, then M is a type II₁ factor if:

1. The action is free: for every non-identity γ belongs to Γ, the measure of the set of points of X fixed by γ is zero;

2. The action is ergodic: the only subsets of X that are Γ-invariant up to null sets are either themselves null or have full measure, i.e. X has no Γ-invariant subsets.

One of the main problems in the theory of von Neumann algebras is to classify such M up to normal *-- isomorphism in terms of their "group/action data". There are various notions of equivalence for group actions. We will be concerned with one levels of equivalence,i.e. Conjugacy.

Conjugacy

Two goup actions Γ ₁→X₁ and Γ ₂→X₂ are said to be conjugate if there is a group isomorphism:

Γ ₁→ Γ₂

There is one concept which quantum theory shares with classical mechanics and classical electrodynamics;That is, the concept of a mathematical "phase-space." According to this concept, any physical system G is at each instantly associated with a "point" p in a phase-space; this point is supposed to represent the "state" of G, and the "state" of G is supposed to be ascertainable by "maximal" observations. Furthermore, the point p_o associated with G at a time t_o, together with a prescribed mathematical law of propagation, fix the point p_t associated with G at any later time t; this assumption evidently embodies the principle of mathematical causality. Thus in classical mechanics, each point of phase-space corresponds to a choice of n position and n conjugate momentum coordinates—and the law of propagation may be Newton's inverse-square law of attraction. Hence in this case phase-space is a region of ordinary 2ndimensional space. In electrodynamics, the points of phase-space can only be specified after certain functions-such as the electromagnetic and electrostatic potential—are known; hence E is a function-space of infinitely many dimensions. Similarly, in quantum theory the points of phase-space correspond to so-called "wave—functions," and hence phase-space is again a function-space—assumed to be Hilbert.

Von Neumann Algabra-Ergodic Action
Lie Groups

If we look at an object such as a crystal with light whose wavelength is longer than the interatomic distances, we perceive it as a continuous rather than a discrete structure. While we can no longer see the local structure, we can still see the global structure. In applied mathematics, structures which are discrete at the atomic level are modelled by continuous structures; e.g. in the equations of fluid dynamics. The first of these jumps between the continuous and the discrete is justified by the belief that the macroscopic properties of the discrete fluid and those of the continuous model are the same, while the second approximation is justified by the functional analytic methods of computing. We consider structures with symmetry, such as cosmological space-time for which an initial assumption is that the universe is homogeneous This assumption is expressed by saying that the laws of physics are invariant under certain changes of coordinate system. In such structures, the symmetries form a continuous group, typically a Lie group. The space itself is usually a quotient space/homogeneous space of the Lie group. Our basic question is then how a homogenous space can be naturally approximated by a discrete structure. In particular, we ask how Lie groups can be approximated by discrete structures, and in this case the natural discrete structures are the discrete subgroups. In some sense, we are concerned with the problem of seeing the global properties of graphs which embed nicely in manifolds, because the lattices of the title are discrete groups, and to a discrete group we can associate its Cayley graph. Alternatively, we are looking for ways of comparing the global geometry of a Lie group with that of a discrete subgroup. A lattice Γ in a Lie group G always assumed connected, and usually algebraic is a discrete subgroup of G such that G/Γ has finite G-invariant measure. For this to become likely, G should be unimodular.

We say Γ is uniform or compact if G/Γ is compact. Two lattices Γ_1 and Γ_2 in G are called commensurable if $\Gamma_1\Gamma_2$ is of finite index in both. One would expect commensurable lattices to have similar global structure. we outline some of the basic algebraic results about lattices in Lie groups obtained roughly between 1955 and 1975, and sketch some of the more recent analytic developments of the theory. In the second part, we describe in more detail the work of U.Haagerup, of Cowling and Haagerup, and of Cowling and R.J. Zimmer on von Neumann algebraic and ergodic theoretic rigidity for lattices. This work is based on the notion of a completely bounded operator in the von Neumann algebra of a group or of an ergodic action.

The QND

The Quantum non-demolition measurement

In 1975, physicists trying to construct a gravitational wave detection antenna faced a dilemma. To detect a gravitational wave with a reasonable probability, they needed to improve the measurement precision for a free mass detector or mechanical harmonic oscillator detector. However, the theoretical analyses for various measurement schemes showed that the precision of such measurement schemes couldn't exceed the standard quantum limits which are imposed by the Heisenberg uncertainty relations. Soon after it was recognized that a certain nonstandard measurement scheme, based on a carefully chosen observable and carefully prepared measuring device, can indeed exceed the standard quantum limits.

Cavity Mirror

Simplest QND measurement scheme of the photon number is the one using a movable cavity mirror. At each time a photon is reflected by the mirror, the momentum 2hk is imposed on the mirror. The mirror's inertia is large enough, so the mirror does not respond to the electromagnetic frequency. It is known that the photon number is preserved when the cavity is deformed. The mirror acquires the momentum from the electromagnetic field during the time interval T. In order to measure this momentum change, we can measure the position of the mirror with a time interval T.

CV QND

The act of measuring a quantum system to acquire information about it, might disturb the system.

Quantum non-demolition;QND measurements allow for the measurement of an observable of a quantum system without introducing a back-action on this observable due to the measurement itself. QND measurements explore the fundamental limitations of measurement and may prove useful in gravity wave detection,telecommunications and quantum control. The traditional domain of experimental QND measurements is continuous-variable CV quantum optics. CV QND measurements are performed using only Gaussian states (those states of the electromagnetic field with a Gaussian Wigner function), working with quadrature components of the field proportional to number and phase in a linearised regime. They have been characterized by considering the signal to noise transfer and conditional variances between various combinations of the input, output and measurement output of the device. These are known as T-V. In contrast, discrete variable quantum optics typically deals with two level quantum systems such as the polarization states of single photons.

Quantum bits or "qubits" can be carried by such systems. Progress in the field of quantum information, in particular in the realization of two qubit gates, has opened a new domain in which QND measurements can be demonstrated. Indeed QND measurements are critical to many key quantum information protocols, such as error correction, and enable new computation models. Until recently it was only in the domain of cavity quantum electrodynamics that interactions sufficiently strong as to probe the qubit domain could be achieved with optical fields. However the work of Knill, Laflamme and Milburn introduced the technique of measurement induced non-linearities and led to proposals for non-deterministic realizations of QND measurements for traveling fields; In these schemes the non-linearity is induced through photon counting measurements. In this paper we investigate the character, characterization, an optical implementation and a fundamental application of QND measurements on qubits. We begin in the next section by describing the basic features that a QND measurement should display. We then propose quantitative measures by which the quality of any QND measurement can be assessed. We consider qubit systems primarily but also discuss the application of these measures to systems of any dimension. Then we consider the trade-off between the accuracy of the

QND measurement and its inevitable back-action on the conjugate observable to that being measured.

Fidelity Measure For QND Measurement

A measurement device takes a quantum system in an input state, described in general by the density matrix ρ, and via an interaction yields a classical measurement outcome,i, of some particular observable. The quantum system is left in the corresponding output state ρ i. To be considered a QND measurement, the device should satisfy the criterion:

" The measurement result should be correctly correlated with the state of the input; e.g., if the input state is an eigenstate of the observable being measured, then in an ideal QND measurement the measurement outcome corresponding to this eigenstate should occur with certainty."

The QND measurement can be tested relative to the criterion by performing repeated measurements of a set of known signal input states $\{\rho\}$. Let $\{|\omega i>, i=1,...,d\}$ be a basis of eigenstates of the measurement of a system with dimension d. The relevant probability distribution: p in of the signal input, which consist of the diagonal elements of the signal input density matrix ρp in$= |\rho,\omega i>$ in the basis of eigenstates of the measurement.

Probability Space

Margenau-Hill Distribution

The Margenau-Hill distribution is the real part of the Kirkwood distributions, so that: $Re(pn)\omega = \int dp\, pn\, M(q,p)/|\rho>q$ where $M(q, p)$ is the Margenau-Hill distribution. $Re[(pn)\omega]$ is a conditional moment of the Margenau-Hill during the short time of interaction. The equation of motion is given by the classical Liouville theorem.

Kirkwood and Wigner Distribution Functions

There exists a great variety of quantum distribution functions in probability space that are widely used in many branches of quantum physics. The Kirkwood distribution function turned out to be a generating function for almost all of them. It is also known as Terletsky or Rihaczek quasi-probability. In quantum physics the knowledge of distribution functions of systems is a very important task. These are functions defined in the phase space of the system, i.e. their definition area belongs to the combined configuration position and momentum spaces (x,p). They are considered as quasi-probabilities since some of them are not strictly non-negative functions in phase space. It has been proved that all of the distribution functions for quantum systems like bosons and fermions can be obtained from one basic quantum distribution function, namely the distribution function first introduced by Kirkwood via the action of suitable convolution pseudo-differential operators. Thus, it turns out that it is important to know the Kirkwood distribution function for the existing different quantum systems.

The basic systems regarded in quantum mechanics are: a particle in a potential well, a harmonic oscillator and hydrogen atom. The main stream of investigations in the literature is focused on Wigner distribution function. The goal for investigations of Kirkwood function is inspired by the fact that the distribution occurs in a very natural way in the quantum field invariants such as energy momentum and spin tensors of the corresponding fields.

Lüders Rule

The Lüders rule describes a change of the state of a quantum system under a selective measurement: if an observable A, with eigenvalues ai and associated eigen projections Pi, → i=1,2;:::, is measured on the system in a state T, then the state transforms to Tk:=PkTPk/tr [Tpk, on the condition that the result ak was obtained. This rule was formulated by Gerhart Lüders as an elaboration of the work of Von Neumann on the measurement process and it is an expression of the projection postulate, or the collapse of the wave function. From the perspective of quantum measurement theory, the Lüders rule characterizes just one albeit distinguished form of state change that may occur in appropriately designed measurements of a given observable with a discrete spectrum. In general, the notion of instrument is used to describe the state changes of a system under a measurement, whether selective or not. The Lüders instrument IL consists of the operations ILX of the form ILX(T)=∑PiTPi, and it is characterized as a repeatable, ideal, non degenerate measurement. In such a measurement, with no selection or reading of the result, the state of the system undergoes the transformation T $I_{LR}(T)=\sum P_i T P_i=\sum tr[TP_i, T_i$, the projection postulate then saying that if a_k is the actual measurement result, this state collapses to T_k. Lüders measurements offer an important characterization of the compatibility of observables A;B with discrete spectra: A and B commute if and only if the expectation value of B is not changed by a nonselective Lüders operation of A in any state T. This result is the basis for the axiom of local commutativity in relativistic quantum field theory: the mutual commutativity of observables from local algebras associated with two spacelike separated regions of space-time ensures, and is necessitated by, the impossibility of influencing the outcomes of measurements in one region through nonselective measurements performed in the other region. The Lüders rule is directly related to the notion of conditional probability in quantum mechanics, conditioning with respect to a single event. According to Gleason's theorem, the generalized probability measures on the projection lattice P(H) of a complex Hilbert space H with dimension dim(H)>3 are uniquely determined by the state operators through the formula $\mu(P) = tr[TP$, for all P in P(H). For any μ and for any P such that $\mu(P) = \xi$, there is a unique general probability measure μP with the property: for all R in P(H), R>P, $\mu P(R)=\mu(R)/\mu(P)$. The state operator defining μP is given by the Lüders form: if μ is determined by the state T, then μ is determined by the state $PTP/tr[TP$.

The Lüders rule is also an essential structural element in

axiomatic reconstructions of quantum mechanics. It occurs in various disguised forms as an axiom in quantum logic; for example, it plays a role in the formulation of the covering law. The Lüders rule has a natural generalization to measurements with a discrete set of outcomes a1,a2;::::, represented by a positive operator measure such that each ai is associated with a positive operator Ai. The general Lüders instrument, defined via the operations T ILX(T) =A1/2 i TA1/2 i, is known to have approximate repeatability and ideality properties. The Lüders theorem extends to general measurements under certain additional assumptions. The Lüders rule is widely used as a practical tool for the effective modeling of experiments with quantum systems undergoing periods of free evolution separated by iterated measurements. It is success-fully applied in the quantum jump approach. The single- and double-slit experiments with individual quantum objects are the classic illustrations of the physical relevance of the Lüders.

Paul Busch,Pekka Lahti

Phase-space representations

We take the overlap $C(t)-<\phi|\omega(t)>$ between the time-evolved state $|\omega(t)>$ and the entangled reference state $|\phi>=\Sigma|\phi a>|a>$. Then the Wigner function provides us with the overlap of two pure states $<\phi|\omega>^2$.

It involves the product of the corresponding Wigner functions integrated over phase space, spanned by the position x and momentum p of the wave function. For the Moyal function of the two states, Equation $<\phi|\omega>-\int dx \int dp W |\omega><\phi| (x.p)$, would be considered as a quite powerful expression since it yields the explicit formula for the scalar product of two entangled states $|\omega>$ and $|\phi>$ of the form $|\phi>=\sum|\phi a>$ $|a>$, in terms of Moyal functions of their oscillator parts $|\phi a>$ and $|\omega a>$

The approach introduced by Michael Berry

A state whose ket is $|n>$ evolves adiabatic under the influence of external parameters that we shall denote collectively as R. These parameters vary on a time scale that is slow relative to the one on which the particle dynamics associated with $|n>$ transpire. As R undergo slow evolution, the time independent Schrödinger equation yields eigenvalues and eigenfunctions at each instant of time, like a series of snapshots: $H(R)|n(R)>=En(R)|n(R)>$ (1

Equation (1) yields a separate set of eigenvalues/eigenkets for each value of R. The standard procedure of electronic structure theory is to solve eqn (1) at different R. Consequently, this results in there being no relationship between the phases of the solutions at different R. We are inclined to think that these unknown phases, if they vary, do so smoothly with respect to R. However, eqation (1 does not say anything about relative phase. After all, it is solved for one value of R, then for another, and so on. In addition, we might want $|n(R)>$ to be single valued and differentiable (except perhaps at the point of closing a circuit), because it will be necessary to take derivatives with respect to the parameters that comprise R as a path in Rspace is followed. To have quantum mechanics, the system should satisfy a

Schrödinger equation, or something close to it, at all points in time throughout the adiabatic evolution:

$$H(R)|\omega n> = id/dt|\omega n> \quad (2$$

The use of a total derivative on the right and side is necessary in order to account for the adiabatic evolution. It turns out that the geometric phase does not depend on the amount of time required to complete the adiabatic cycle. However, it is necessary to take into account the changes that transpire in the space of parameters R, and this is the reason for using the total derivative instead of a partial derivative in eqation (2. The state vector $|\omega n>$ includes phase factors for the usual Schrödinger eigenstate phase, as well as an additional phase $pn(R)$. This phase is due to the adiabatic evolution of R. The rest is standard quantum mechanics. The term pn should be present; It is an admission of our ignorance regarding how the phase evolves with R. Because phase evolves on the parameter space, we see that only R affects the phase. The phase varies according to changes in R, not how long it takes these changes to happen. An equation for pn is obtained by putting the $|\omega n>$ into the right hand side of eqation (2. This yields three terms on the right hand side, one of which cancels the term on the left hand side. Time does not play an explicit role in determining the phase. Namely, pn depends on the path, not how long it takes to traverse it. Different paths that end at the same point in general give different phases, despite the fact that the time elapsed in traversing them is the same. Thus, pn can not be written as an explicit function of time. Using different amounts of time to traverse the same path yields the same phase. Said differently, for a given path, the phase accumulated in going from Ro to R is the same whether passage is carried out slowly or less slowly, as long as it is done adiabatic. If R returns to its initial value via a closed path C, the geometric phase $\rho n(C)$ is that of a completed circuit. If $|n>$ is single valued its differentiation can be carried out with impunity along C. We will see that $|n>$ can be assigned an arbitrary, parameter dependent phase without changing $\rho n(C)$, ensuring that a single valued wave function can be used. Even if $|n>$ begins life not single valued, it can be made single valued, including the point at which the circuit closes. In 1984, Berry pointed out that in a cyclic adiabatic process, that is one in which the slowly time varying Hamiltonian returns to its original form via a circuit C, a quantum state might acquire a geometrical phase factor in addition to the normal dynamical phase factor. In an elegant calculation, Berry showed that if the circuit occurs in the vicinity of a degeneracy of the Hamiltonian in parameter space, then the geometrical phase is proportional to the solid angle Ω subtended by the circuit at the degeneracy.

As an illustrative example, He considered spins in a magnetic field characterized by slowly varying parameters R. The Hamiltonian for this system has a degeneracy at R=0 where B=0. For the simplest case of a cone, θ constant, the solid angle is $\Omega=2\pi(1-\cos\theta)$. Imagine that such a conical circuit is traversed adiabatical, that is to say with small σ, where $\sigma=2\pi/T$ and T is the period of the circuit. A spin eigenstate with magnetic quantum number m should accumulate a geometrical phase $\sigma(C)=2\pi$ m$(1-\cos\theta)$ in addition to the dynamical phase. Wilczek and co-workers and Cina have suggested that a manifestation of the geometrical phase should be observed in interference between eigenstates, for example in the evolution of a coherent superposition of states m and m'. Such a superposition corresponds to magnetization or to higher rank tensor coherences and the phase changes of such coherences have been observed for states in N.M.R. undergoing non-adiabatic circuits. Upon completion of an adiabatic circuit, a coherence should acquire a geometrical phase change or extra rotation, in addition to the dynamical procession angle ϕd. Chiao reported a classical optical version of Berry's experiment for spins in a magnetic field in which the plane of linearly polarized light which corresponds to a superposition of the m=+-1 photon states, was rotated by a geometrical phase imposed by helical wound optical fibers. Tycko performed a nuclear quadru pole resonance experiment in which the geometric phase of a spin-3/2 was observed during rotation of a crystal, thereby moving the quantization axis of the electric field gradient in a cone. The geometrical phase is also related to early work on fractional quantum numbers in molecules and the classical work on conical intersections by Herzberg and Longuet-Higgins.

Choosing Phase

There are different ways to obtain $\rho(C)$; For example, if the integrand vanishes along C, $\rho(C)$ can be calculated at the close of the circuit where the wave function is discontinuous. The integral vanishing along C implies ω is orthogonal to $d\omega$ which is not surprising. At $\alpha= 2\pi$, ω undergo a phase change of π.

Thus, the equivalence of the approaches has been demonstrated. The geometric phase value of $-\pi$ has been obtained by integration on C both with and without a single valued wave function. Suffice it to say that α depends on the Hamiltonian in a manner such that as α moves toward 2π the system moves around a conical intersection. The ½ intrinsic angular momentum would arise as to the ω degeneracy.

To verify that the transformed wave function is single valued, multiply ω_1 in the following Max. $\omega_1 = [\cos\alpha/2 \ \sin\alpha/2[\phi_1 \ , \omega_2 = [-\sin\alpha/2 \ \cos\alpha/2[\phi_2$ by $e i\alpha/2$ and write $\cos\alpha/2$ and $\sin\alpha/2$ in terms of exponentials.

Spin and Roothaan–Hall

The electron is known to have a spin with a quantum number $s=1/2$, the z-component of which is quantised to take one of the possible values, $m_s=\pm 1/2$. In order for electron spin to have meaning in Hartree–Fock theory, we should let the orbitals depend on it.

The high pressure properties of Krypton

The importance of the four-body contribution in compressed solid krypton was first evaluated using the many-body expansion method and the coupled cluster theory with full single and double excitations plus perturbative treatment of triples. All different four-atom clusters existing in the first- and second-nearest neighbor shells of face-centered cubic krypton were considered, and both self-consistent-field Hartree-Fock and correlation parts of the four-body interaction were accurately determined from the ambient conditions up to eight-fold volume compression. We find that the four-body interaction energy is negative at compression ratio lower than 2, where the dispersive forces play a dominant role. With increasing the compression, the four-body contribution becomes repulsive and significantly cancels the over-softening effects of the three-body potential. The obtained equation of state EOS was compared with the experiments and the density-functional theory calculations. It shows that combination of the four-body effects with two- and three-body interactions leads to an excellent agreement with EOS measurements throughout the whole experiment.

Recently, the theoretical studies have revealed that the use of a two-body potential together with the three -body correction could result in a good prediction of EOS in dozens of Giga pascal pressure range, however with less satisfactory agreement at higher pressures. It was suggested that high order many-body contributions, e.g. four-body system might be of some importance for highly compressed solids of heavy rare gas elements. The two-body potentials of Krypton atoms may have been well established in the previous observations, and the three-body interactions of krypton were investigated too in the works of Loubeyre,Barker and Freiman, though few studies on the four-body contribution in krypton and other rare-gas elements have been conducted. Due to the weakness of the dispersive energy of Van der Waals very accurate ab initio methods should be required for calculating the interaction energies of rare gas atoms. In fact the density functional theory

116

DFT may not be accurate sufficiently enough to account for the long-range dispersive interactions. At present, many-body expansion technique with the wave-function based correlation methods is expected to give more accurate description for rare gas solids. Recently, one of the wave-function based correlation methods, the coupled cluster theory with full single and double excitations plus perturbative treatment of triples CCD(T) has been successfully applied to calculate the two-body potentials of rare gas dimers of He_2, Ne_2, Ar_2 and Kr_2 as well as the three-body interaction for neon trimer Ne_3. There is little information about four-body interaction for Krypton and other rare gas elements. The calculations of four-body interaction for rare-gas quadruplet are more difficult rather those of two and three-body interactions as to the rise of computational effort which is proportional to the power of seven of the number of correlated electrons at CCSD(T) level; and for Krypton atom the calculations are much more demanding rather those of the light rare gases which could be due to the occupied d-shell electrons. In addition, since a lot of different geometrical quadruplets exist in the crystal lattice, it should be necessary to include a sufficient number of configurations to obtain a reliable interaction energy. Though Rosciszewskiv have investigate the short-range four-body interaction energy in rare gas solids, their observations were restricted to one four-atom cluster of the regular tetrahedron.

Chunling Tian, Na Wu, Fusheng Liu, Surendra K. Saxena Xingrong Zheng

Autoionizing States Of Noble-Gas Atome And Ions

Long-lived autoionizing states of argon, krypton, and xenon atoms, and of singly-charged argon ions, are observed by the mass-spectrometric method. Special ion sources were used in which atoms were excited or ionized by electron impact, and the autoionization process was registered by means of the ions produced in a separate chamber. The long-lived autoionizing states of the atoms lie between the first and second ionization limits, $np5$ $2P3/2$ and $np5$ $2P1/2$. The low ionization rate (~ 10-6 sec) is due to the large values of the principal quantum number n and of the orbital angular momentum of the excited electron. These states are not observed in optical spectra because of the selection rules for dipole transitions.

It is well-known that ions and atoms of the noble gases possess long-lived Rydberg states that lie near the ionization limits and that can easily undergo ionization near a metal surface in an electric field or in collisions with molecules. The atoms and ions of the noble gases possess two or more ionization limits to which different Rydberg series converge. Excited states lying above the lowest ionization limit can autoionize. Photoabsorption experiments have indicated that some of these states have short lifetimes ~ 10-13 sec with respect to autoionization. In these experiments only optically allowed states were excited, to which dipole transitions are allowed by the selection rules; these were ns and nd states. However, electron impacts also excite optically forbidden states such as np and nf, which can possess small probabilities of autoionization. It has been reported that collisions between electrons and noble-gas atoms produce longlived autoionizing states of Ar+ and xe+ ions having lifetimes ~ 10-6 sec, but the nature of these states was not discussed. The given conclusion was based on the observed way in which the ion intensity was influenced by the initial gas pressure and electron energy. The principal experimental results obtained are that 1) in the process A+ A2+ +e (1) the intensity of the produced doubly-charged ions A2+ is directly proportional to the intensity of the initial singly-charged ions A+, 2) the doubly-charged ions A2 + are formed from the singly-charged ions in accordance with (1) when the electron energy slightly surpasses (by about 0.5-1 eV) the threshold for the production of doubly-charged ions from atoms in the process A+ e +A2+ + 3e   (2) These results would provide a sufficient basis for the conclusion that A2+ ions result from the autoionization. We shall here consider only the lowest ionization limits, which are formed by the removal of an outer np6 electron from an atom or an outer np5 electron from a ion. of A+ ions if the initial A+ ions did not possess additional highly-excited states that converge to the first ionization limit and are therefore unable to autoionize although their excitation energies are close to autoionizing states. However, the foregoing discussion points to the existence of such states, which sometimes behave like autoionizing states. For example, the ionization processes of these highly excited states when ions in an electric field collide with residual gas molecules (forming a background), or occurring near the metallic surfaces of collimating slits, are also directly proportional to the intensity of the initial A+ ions. Nonuniformity of the electron beam energy and a large difference (by a factor of about 10^5) between the intensities of the A+ ions and the A2+ ions produced therefrom according to (1) make it difficult to determine the exact small difference between their thresholds. When the ion beams passed through narrow slits, It was likely that a considerable fraction of the A2+ ions, which were attributed only to autoionization of A+, had resulted from the ionization of highly excited A+* ions; these could include ions in autoionizing states, near the metallic surfaces of the slits. It is therefore necessary to study long-lived autoionizing states of ions under more determinate experimental conditions. Moreover, if singly-charged noblegas ions should appear in such states, the same should also apply to the corresponding atoms. The present work is an investigation of these effects.

EXPERIMENT

In the mass-spectrometric investigation of long-lived autoionizing states our basic task is the discrimination of these states from lower-lying long-lived highly excited states. The task is difficult because the two types of states have very close energies and, as already mentioned, in some experiments they behave alike. Our work was done with a mass spectrometer and a esp. designed ion source that enabled us to discriminate these two kinds of states. Autoionizing States of the Atoms The two-chamber ion source had been changed in the following way, The second chamber Ka was lengthened to 13 mm in order to permit a larger number of decays; its entrance and exit slits were enlarged to 4 x 16 and 6 x 16 mm, respectively, so that the atom beam collimated by the slit S_1 = 1 x 8 mm would not come into contact with the edges of the chamber slits. The narrow slit S_a = 1 x 9 mm behind the second chamber was covered with a ""70% transparent fine copper grid. This was followed by the exit slit of the ion source. The ions were retained in the first chamber by the retarding field V_R =50 V, the field V_c =50 V of the deflecting condenser, and a ""250-gauss magnetic field that also collimated the electron beam. A 3.05-kV accelerating potential was applied between the slit S_1 and the electrode A; the potential at the second chamber was 2.8 kV, and that at the slit S_a with the grid was 2. 5 kV. These quantities could be regulated in order to study the influence of the electric fields on the ionization of the highly-excited atoms. The gas pressure in the source region was usually,.., 10-5 10-4 Torr. The ion current was registered with a U_{1-2} electrometric amplifier.

Autoionizing States of the Ions

We investigated only singly-charged argon ions, which are the most suitable for the present purpose. Long-lived autoionizing states were observed with a mass spectrometer and a special ion source. Our source enabled us to distinguish between doubly charged Ar_{2+} ions resulting from the autoionization of Ar_+ * and other Ar_2 + ions resulting from different processes, such as ionization at the edges of the collimating slits or ionization in collisions with Ar and background atoms. The source functioned as follows. Ar_+ ions accelerated to 2.8 keV entered chamber K_2, to which a certain potential ± V K was applied. The doubly-charged ions formed in this chamber from singly-charged ions received the following energy: 2.8 'f VK (the energy of the singly-charged ions entering the chamber) plus ± 2VK (received by doubly-charged ions leaving the chamber); the total was 2.8 ± VK. The peak representing these ions could therefore be shifted along the mass scale to a position that is free of overlapping peaks. This peak could be formed by Ar^{2+} ions produced from Ar^{+*} ions by only two processes-autoionization and stripping when argon atoms collide with residual gas molecules. Ionization of highly excited Ar^{+*} at the edges of the chamber slits was excluded because these slits were very much wider (23 x 23 mm) than the nearby collimating slits S_1 (1 x 7 mm) and S_2 (10 x 10 mm).

The dimensions of the remaining slits were 1 x 4 mm. Chamber K_2 was 212 mm long. In the first chamber K_1 highly excited ions are produced by collisions between electrons (E) and atoms (A). These ions are extracted (B), focused (F) and directed into the second chamber K_2, where autoionization occurs. Ar^{+*} ions that passed through the grid lost an electron and were thus converted into Ar^{2+} ions that were registered on the mass scale in accordance with the potential applied to the grid. The described apparatus was used to study how the intensities of Ar^+ produced in chamber K_2 by collisions of electrons with argon atoms, and of Ar^{2+} produced from singly charged ions traversing the collimating slits, grid, and chamber K_2, were influenced by the following factors: Ar pressure, the background, and the electron current and energy. The background pressure was varied by closing valves to reduce the rate of evacuation.

RESULTS

Our results indicate that multielectron atoms and ions of noble gases exist in long-lived autoionizing states having lifetimes ~ 10-6 sec and lying close to the corresponding ionization limits. In the case of noble-gas atoms these states lie between the two ionization limits 2P3/2 and 2P1/2 to which respective Rydberg series converge. These states result from the excitation of only a single atomic electron. The ground level of doubly-charged noble-gas ions is a triplet; the three sub-levels 3P0, 3P1, and 3P2 of Ara.- are 43.58, 43.53, and 43.93 eV, respectively. These ions will therefore possess Rydberg series converging to these three ionization limits. The levels that converge to the lowest limit 3P2, are highly excited long-lived states. The levels lying above the first ionization limit will autoionize. Two spontaneous decay modes are possible for an excited autoionizing state:

1) a non radiative transition into the continuous spectrum (i.e., autoionization),

2) a radiative transition to a lower level or the ground state.

For an allowed transition of the first kind, induced by inter electronic Coulomb interaction, the transition rate is "'10-15-10-14 sec while for allowed transitions of the second kind it is "' 10-9 sec. In the cases of the excited atoms and ions that we registered both transitions must be forbidden, or their probabilities are considerably reduced. This follows directly from the experimental result that autoionization of the excited atoms and ions was observed ,..10-6 sec after their formation. For a radiative transition the mean lifetime Tn of a highly excited hydrogen-like state is dependent on n: T n "' 10-9 n4,5, Therefore with n> 6 the lifetime will be sufficiently long to permit observation of the described states in our experimental work. The situation is different with regard to autoionization, which can result from either Coulomb or magnetic interaction between electrons. In both instances the conservation of parity and of the total angular momentum comprise strict selection rules. For Coulomb autoionization in the case of LS coupling it is also necessary to conserve the total orbital angular momentum L and the total spin S; However, these are not strict rules and can be violated as a result of magnetic interactions. For the noble-gas states of present interest it is known that LS coupling breaks down because of strong spin-orbit interactions of atomic core electrons. Then jl, or possibly jj, coupling is the more suitable type. The approximate selection rules for autoionization through Coulomb interaction will here be different and will be derived by a calculation that is outside the scope of the present article. We note only that the strict selection rules (conservation of total angular momentum and parity) hold true for the other types of coupling also. The considered states of the discrete spectrum lie above the first ionization limit

2P3/2 and the total angular momentum of their atomic core is J = 1 /2 2P1/2); Thus, as a result of autoionization the total angular momentum of the atomic core should undergo a unit change (1/2

3/2). This transition can occur only through Coulomb or magnetic interactions between core electrons and excited outer electrons, but not through interactions between electrons within the core. We are thankful to B. Firsov, S. I. Grishanovaya, V. S. Senashenko, and B. M. Smirnov for discussions.

Hartree-Fock-Roothaan theory

The Hartree–Fock–Roothaan (HFR) or basis-set expansion method is a convenient and powerful tool for the study of electronic structure of atom. The most frequently used basis functions for atomic calculations are Slater type orbitals STO defined as: $\sum$n lm(ζ,r)=[(2ζ)n +1/2 /[Γ (2n +1)[^{2} rn$\ast$ $-$ 1 e $-$ ζr Slm(θ, ϕ);

where Γ is the Gamma Identity and ζ considered to be Riemann zeta function.

Spin-Extended Hartree-Fock Method

The partial filling of degenerate one-electron levels calls for the use of the well-known Roothaan scheme for open shells in atomic calculations. This method takes into account electron correlation and provides the correct symmetry of the wave function even in the presence of degeneracy. When there is no degeneracy at place one set of orbial should be left, whereas

125

there is degeneracy there would be no pairing and we have two set of orbitals. In the restricted Hartree-Fock, a one-determinant wave function is constructed from orbital pairs, whilst in the presence of degeneracy which requires different set of orbitals for different spins, the wave function is constructed from a twice as large set of the spatial orbitals which appear with spins ß and α. The partial location of the ß and α electrons results in lowering of the symmetry of the spin-splitting matrix with maintenance of the regular D4nh symmetry of the charge-density matrix as well as the matrix z^2. In Spin-Extended Hartree-Fock Method the last two matrix define quantities having physical meaning, while the odd powers of z appear only in the generalized Fock operators Fα and Fß, which thus also have a lowered symmetry. The arrangement and filling of the levels of Fß would be accomplished by the removal of the degeneracy of the non-bonding levels as a result of the lowering of the symmetry. Their corresponding orbitals also transform according to the representations of the D2nh group along with the complete wave function. In the restricted Hartree-Fock the filling of the highest degenerate level is equivocal. By applying the method which requires different set of orbitals for different spins, the 4π spin up electrons fill the levels of Fß operator. We might be able to demonstrate this by applying the operation of rotation through an angle equal to $2\pi/4n$ to the product of the rotation, which corresponds to certain columns of the unprojected determinant. We might also be able to prove by some relatively simple transformations that ß and α spin inversion occurs in it in addition to the significant transposition of the multipliers, when the arguments being put in a definite sequence.

After the spin projection, which is inherent in its essence to the method using different set of orbitals for deifferent spins and also symmetrizes the wave function with respect to ß and α, the result of the application of the rotation operation is no longer accompanied by the undesirable ß and α inversions.

In fact this should be clear from the property of the spin-projection coefficients for zero spin, where the subscript indicates the number of transpositions of ß and α. As a result the total wave function transforms in accordance with an one dimensional representation of group D4nh, which can be specified from the form of the degenerate non-bonding set of orbitals Xn and θn. Though the expressions of the projected version of this method for Fß-Fα appear to be considerably more complicated even in their compact matrix form, it should not be forgotten that here, in contrast to the approaches of the Roothaan type, electron correlation is taken into account. Unlike the restricted Hartree-Fock this method is capable of describing the configuration of structures with both unequal and equal alternating bond lengths. The energies of the π-electronic conjugation decrease both in this method and in the RHF, as the degree of alteration k which might be simulated by the change in the resonance integral of the neighboring bonds, is increased. The total energy has a lower limit owing to the deformation of the σ component. While in the RHF the difference between the orders of the neighboring bonds ΔP tends to a constant value, in AMO method the bond orders alternate to a considerably lesser extent, tending to the same limit as k 1; Moreover, the decrease in the energies of the π-electronic conjugation occurs appreciably more slowly in the AMO and thus here the consideration of electron correlation apparently favors weaker alternation of the bond lengths. In the view of the increase in ΔE as k 1, the role of electron correlation is esp. significant in the case of configurations with weakly alternating or equal bond lengths.

Roothaan Parameters

In open-shell calculations within the restricted Hartree-Fock (ROHF), the coupling between the closed and the open shells could be specified using two parameters α and β, which depend on the type of the open shell, the number of electrons in it (the electron configuration), but also on the state to be calculated. For example, there are three states arising from the s2p2 configuration of an atom (3P, 1D, 1S) which have different values of α and β. Note that not all open shell systems can be handled in this way. It is possible to specify α and β for atomic calculations with sn, pn, d1 and d9 configurations and for calculations on linear molecules with πn and δn configurations, Furthermore, it is possible to do calculations on systems with half-filled shells where $\alpha=1$, $\beta=2$.

Spin functions β and α

We let the orbitals depend on the spin by attaching an extra label (superscript) to it, $\phi k\alpha=\phi\alpha k(r)\alpha$, $\phi k\beta=\phi\beta k(r)\beta$, The functions $\phi k\alpha$ and $\phi k\beta$ are spin orbitals that depend on spatial and spin coordinates. Having attached a spin label to the orbitals, we should write sums of the orbitals in the following manner: without spin : $\sum$ k-1 n $\Rightarrow$ with spin: $\sum$ σ- α,β X k-1 nσ (nα+nβ=n) There is a sum over σ over the two spin cases and then (dependent on the first sum) a second sum over the number of orbitals with that spin.

Gamma Function Identitiy

There are a number of rotational conventions in common use for indication of a power of a gamma functions. While authors such as Watson (1939) use Γn(z) (i.e., using a trigonometric function-like convention), it is also common to write [Γ(z)]n. The gamma function can be defined as a definite integral for R[z]>0 (Euler's integral form)

$\Gamma(z) = \int t(z-1)e(-t)dt$ (1

$= 2 \int e(-t^2)t(2z-1)dt,$ (2

Or

$\Gamma(z) = \int 0 \rightarrow 1[\ln(1/t)]^{\wedge}(z-1)dt.$ (3

The relationship between Γ(z) and the Riemann ζ(z) function ζ(z) is given

$\zeta(z)\,\Gamma(z) = \int[(u(z-1)/(eu-1)[\,du;\quad$ For R[z]>1   (4

Zeta Function

The application of these new data is illustrated through evaluation of the main minimum of T(x) to 31D.

The zeta function As an analytical continuation has the functional equation of the form ζ(s)-x(s) ζ(1-s).

In what follows we are going to use: $x(s) = \pi^{\wedge} s - 1/2\ \Gamma(1-s/2)\ /$

$\Gamma(s/2)$ (2

where $\Gamma(s)$ denoted the Gamma function. Recall that the zeta function has no zeros in the region $1<\delta$, the functional equation (1.5) reveals that the only zeros for $\delta<0$ are given by $x(s)$ or precisely by the poles of $\Gamma(s/2)$.

The poles are located at s= 0; 2; 4;:::: and describe the so called trivial zeros of ζ at s_ -2k. At s-0, we find a removable singularity, since the pole of Γ compensated by the pole 1-s in eq.1. Hence, all other zeros have to lie in the critical strip $0<\delta$ <1. Due to the functional equation (1), they should be located in symmetry to the axis δ-1/2, the critical line.

Riemann Hypothesis

The Riemann Hypothesis states that non-trivial zeros of the zeta function have real part $\sigma=1/2$; There has been considerable excitement about the connection between the Riemann Hypothesis and quantum mechanics. Finding a proof has been the holy grail of number theory since He first published his hypothesis. It was identified by Hilbert in 1900 as one of his 22 mathematical challenges for the 21th Century, and by the Clay Mathematics Institute in 2000 as one of its seven $1million Millennium Prize Problems. From a conference in 1996 in Seattle, aimed at fostering collaboration between physicists and number theorists, came early evidence showing there are striking similarities between the Riemann zeros and the quantum energy levels of classic chaotic systems. There are certain attributes of the Riemann zeta function called its moments which should give rise to a sequence of numbers,

However, for the past century only two of these moments were known: First, calculated by Hardy and Littlewood in 1918; and the Second, calculated by Ingham in 1926. While the number theorists tried to do the same using their methods, Prof Jon Keating and Dr Nina Snaith at Bristol describe the energy levels in quantum systems using random matrix theory. Using RMT methods they produced a formula for calculating all of the moments of the Riemann zeta function.

Riemann zeta function implication

Proving the Riemann hypothesis is to give a spectral interpretation of the trivial zeros; if the trivial zeros can be interpreted as the eigenvalues of $1/2 + iT$, where T is a Hermitian operator on some Hilbert space, then since the eigenvalues of a Hermitian operator are real, the Riemann hypothesis follows. The best evidence for the spectral interpretation comes from the theory of the Gaussian Unitary Ensemble GUE, which show that the local behavior of the trivial zeros mimics that of a random Hamiltonian. Gutzwiller gave a trace formula in the setting of quantum chaos which relates the classical and quantum mechanical pictures.

Given a chaotic classical dynamical system, there will exist a dense set of periodic orbits, and one side of the trace formula will be a sum over the lengths of these orbits. On the other side will be a sum over the eigenvalues of the Hamiltonian in the quantum-mechanic analog of the given classical dynamic. This setup resembles the explicit formulas of prime number theory. In this analogy, the lengths of the prime periodic orbits play the role of the rational primes, while the eigenvalues of the Hamiltonian play the role of the trivial zeros of the zeta function. Based on this analogy and pearls mined from Odlyzko's numerical evidence, Sir Michael Berry proposes that there exists a classical dynamical system, asymmetric with respect to time reversal, the lengths of whose periodic orbits correspond to the rational primes, and whose quantum-mechanical analog has a Hamiltonian with trivial zeros equal to the imaginary parts of the nontrivial zeros of the zeta function.

In the Riemann hypothesis We can show the difference between the trivial and non-trivial zeros of the ζ function by plotting their values in the complex plane, which is a coordinate plane resembling our ordinary xy-plane. However, instead of the x-axis, we have the real axis; instead of the y-axis, we have the imaginary axis. When we graph the zeros of ζ in the complex plane, the trivial zeros are all real, so they lie on the real axis (the horizontal line). The other points on the graph are the non-trivial zeros of the ζ function; if the Riemann Hypothesis is true, then they should have real part. In other words, they should lie on the critical line, the vertical line $\sigma=1/2$. Note that the Riemann Hypothesis only concerns itself with non-trivial zeros. As a result, by using the term zeros of the ζ function we refer to non-trivial zeros, unless otherwise stated. To approch this F.Carlson introduced the function $N(\beta,T$, which equals the number of zeros of $\zeta(s)$ in the rectangle $\beta \leq \delta < 1$, $0 < t \leq T$. A.E. Ingham proved that if $N(\beta,T)=O(Tb(1-\beta) \log B T$ holds uniformly for $1/2 < \beta < 1$, then $pi(x+x\theta)-pi(x) \sim x\theta/\log x$ for $1/b < \theta$.

The Riemann hypothesis implication for the Zeta function

The zeta-function on the critical line δ-1/2; has the most sure asymptotic for the discrete moment: $\sum m \leq M$ $\zeta(1/2+iCm)$; Turan says that the behavior of $\zeta(s,$ is inextricably connected with the distribution of primes. Therefore he proves $N(ß,T, \leq T^2(1-ß$ [exp(13 log 0.18 T, under a hypothesis that has nothing to do with primes. Laszlo Kalmar called the following statement the quasi-Riemann hypothesis: One can find a number $1/2 \leq ß<1$ such that $\zeta(s,$ has only finitely many zeros in the half plane δ-1/2. He gave a necessary and sufficient condition for the quasi-Riemann hypothesis to hold.

The Hamiltonian Potential

There is nothing more repulsive that the harmonic potential that provided equiespaced levels; that is to say: its levels conforms a perfect crystal, perfect repulsion. The fact that we have a log(E) law implies that the repulsion is not perfect but the distance between levels (roots) is 1/log(E), almost constant, esp. at high energies or short wavelengths. This is what switch the physical idea of energies with the mathematical function ζ.

> Asymptotic for δ-1/2 The definition of the Dirichlet series
> It becomes obvious that the behavior of Asymptotic for large positive real parts of s=δ+iτ is governed by the function Fe(s)=1 + e-sln2; (6.1)

Indeed, The only curves which start at δ+1/2, are the outgoing separatices of the points that are located between the incoming ones at τ-+-(2k+1)π/ln2. The flows near the outgoing seperatices point to the right leading back to the critical strip. The incoming separatices with imaginary part larger than π/ln2, cross the complex plane from left to right forming natural groups of non-trivial zeros. In contrast, the outgoing separatices with bigger τ's have to end in a non-trivial zero of the ζ, since they alternate with the incoming separatices. Moreover, it seems that all the curves coming from right do not cross the critical line. However, the separatix at τ-π/ln2 and the flow around it acts differently. Although the separatix comes from the right and can be considered as outgoing part with respect to the hyperbolic point at s'o, it is simultaneously an incoming separatices for the first trivial zero of the ζ'. The derivative ζ' has no zeros in the left half of the critical strip 0<δ<1/2. Hence, it is worthwhile to investigate the Newton flow of ζ and its derivatives, since it would reveal whether the Riemann hypothesis was violated.

The Newton flow of ζ reveals the separatices
The behavior of the Newton flow is governed by the asymptotic. From the eq. $\zeta(s)=[1+e-(1-s)\ln2[x(s);$ (6.2)
follows that the behavior of ζ for δ-1/2 is mainly determined by x in the equation 2. i.e. $x(s)=\pi^{\wedge}s-1/2\ \Gamma(1-s/2)\ /\ \Gamma(s/2)$; hence the speratices are real with phase o π in alternating sequence. They direct the flow into the trivial zeros of ζ' on the negative real axis, which are slightly shifted to the right of the corresponding zeros of x'. Some real positive lines are

134

separatices, the other flows end in the non-trivial zeros of ζ. In the asymptotic for δ-1/2 given by eq. (6.1), Even though separatices are real and positive, only the incoming parts start at negative leading into the zeros of ζ' at δ-1/2 and τ-2πk/ln2.

These form natural groups of the non-trivial zeros of ζ. The corresponding outgoing separatices are the only ones starting at +1/2. Their initial imaginary parts is τ-(2k+1)π/ln2, and they might end in a non-trivial zero of ζ. However, the separatices on the real line and the one starting at 1/2+iπ/ln2 are special.

The separatices of x and ζ

The influence of the Dirichlet series in the approximation eq. (6.2) is quite small in the left half of the complex plane for δ<-4, the difference between the real separatices of x and ζ indicating the phase π is the only visible on the right side of the critical line. In contrast, the separatrices through the non-trivial zero of x on the critical line does not appear as separatix in the flow of ζ, as well as other separatices to the right. There are two real separatices of non-trivial zeros that enclose the flow between the poles s-1 and s-3 and the zeros s-0 and s-2 of x. In contrast, the Riemann zeta function only has one real separatice in this region, since ζ has only one pole at sp-1. This separatice starts at s- +iπ/ln 2 and leads to the first trivial zero of ζ' at δ. The real separatices of ζ can be matched to some of the positive real lines of x, whereas the negative real lines are similar to the negative curves of ζ, which end in a nontrivial zero of ζ.

Non-trivial separatices

So far, we have only described the real separatices of ζ determined by the trivial zeros of ζ' located on the real axis or at σ-1/2. The outgoing parts of the last form natural groups of two or more non-trivial zeros of ζ for τ>----. Hence, the flow into different zeros of one group should be separated by non-real curves through hyperbolic points. This property of the Newton flow indicates that there are n-1 non-trivial zeros of the derivative ζ' in the region of a group with n non-trivial zeros of ζ, provided that the higher derivatives of ζ do not vanish at the hyperbolic points. Otherwise, there are less zeros of ζ'. Yet, should be noted that up to τ-100 no hyperbolic point exists where the second derivative of ζ vanishes.

Combined-quantum systems

Our physical approach towards the Riemann zeta function takes advantage of its similarity to the time evolution of a quantum state. When we choose the associated Hamiltonian appropriately, we can reproduce the phases of the sum m and s in the different representations of ζ. The overlap with an

136

adequate reference state then takes care of the amplitudes.

Theory of Attosecond Transient Absorption Spectroscopy of Krypton for Overlapping Pump +Probe/Pulses

The interaction of matter with light is a key process in physical systems on any length scale. The fundamentals of matter-light interaction can be best studied in atomic systems due to their relative simplicity. The absorption of light promotes electrons into excited states. If enough energy is absorbed by the system, one or more electrons can leave the atom, i.e., ionization takes place. The most common types of ionization are: single-photon and few photon ionizations, above-threshold ionization , and tunnel ionization. Recently, high harmonic generation (HHG) has become a major tool in attosecond physics, allowing one to generate ultrashort light pulses with broad spectral bandwidths. From the ability to generate attosecond pulses, an entire new research area has emerged focusing on electronic dynamics and molecular motion on their fundamental time scale. A particularly interesting aspect is the electron motion and the corresponding hole creation dynamics during the ionization process . The high pulse intensities used in these experiments distort significantly the potential of the electrons such that it is possible for the electron to tunnel through or even travel over the barrier out of the system (i.e., tunnel ionization or barrier suppression regime, respectively). A well-known model to describe tunnel ionization in atomic systems is the Ammosov, Delone, and Krainov (ADK) model. The aim of this study is to investigate the ion population dynamics in krypton within the pump pulse. We show that the instantaneous ionic state population can be well captured by the transient absorption spectrum even for overlapping pump and probe pulses. Furthermore, we observe strong modifications of the absorption lines in the transient absorption spectroscopy when pump and probe pulses have the maximum overlap. We show that these deformations can be understood by relative phase shifts in the ionic dipole. We identify that the highly non-perturbative dressing of the N-electron states (particularly with the neutral ground state) is responsible for the phase shift. Also the dressing of the ionic (N − 1-electron) states, which leads to energy shifts in the ionic states, contributes to the phase shift. The latter one is, however, much weaker than the first dressing mechanism. Note that these two dressing mechanisms are quite different in nature. The first mechanism dresses Nelectron states and the second one dresses N − 1-electron states.

To capture these dressing mechanisms during ionization, a description of the entire N-electron system is required. We describe the dynamics of the full N-body wave function with a time-dependent configuration interaction singles (TDCIS) approach. The description of the pump and the probe steps requires at least two active electrons, since the pump pulse ionizes an outer-valence electron and the probe pulse resonantly excites an inner-shell electron into the generated hole. Therefore, it is crucial to use a multi-channel model, which go beyond the single-active electron (SAE) approximation.

THEORETICAL METHODS

Equations of Motion

By allowing only one electron to get excited or ionized out of the ground state configuration, we strongly reduce the complexity of solving Eq. of motion. A suitable way to achieve this goal is by exploiting the configuration interaction (CI) language and describing the N-body wave function in terms of the Hartree-Fock ground state. This approximation is known as CI-Singles (CIS). The Hamiltonian operator is the residual electron-electron interaction, which go beyond the mean-field potential. The vector and dual vector with respect to the symmetric inner product required because of the non-hermiticity of Hamiltonian operator. The dipole interaction between singly-excited configurations reduces to transitions between states of the excited electron and transitions between ionic states. From the full N-body wave function one can construct the ion density matrix (IDM) by tracing over the excited electron. The CAP is placed far away from the atom such that an electron so far out does not affect the ion, esp. the ionic states. Therefore, the absorption of an electron by the CAP results only in an artificial loss of norm Spin-Orbit Splitting In order to include the effect of spin-orbit splitting in the occupied orbitals we account for spin-orbit splitting with degenerate-state perturbation theory within the (n,l) orbital manifold.

The occupied orbital i is, then, characterized by the quantum numbers, the principal quantum number,the orbital angular momentum, the total angular momentum, and the projection of the total angular momentum onto the polarization direction of the external laser field. The orbital energies ε_i are taken from experimental ionization potentials. For the virtual orbitals, we can neglect spin-orbital splitting and use the quantum numbers n_a, l_a, σ_a, $m_L{}_a$ to classify the orbitals, where σ_a is the spin component in the laser polarization direction and $m_L{}_a$ is the projection of the orbital angular momentum onto the laser polarization direction. After the introduction of spin-orbit splitting for the occupied orbitals, we cannot make use of the σ and m_L symmetries independently to reduce the number of singly-excited configurations; however, not all symmetries are lost and we find that some are up to a global phase invariant under the parity transformation. Transient Absorption for Overlapping Pulses The transient absorption signal is a direct measure of the cross section of the System. The probe pulse was treated in first-order perturbation theory such that it was possible to give an analytic expression for the transient absorption signal as a function of the instantaneous IDM ρIDM(t). The pump pulse, usually a strong-field NIR pulse, which ionizes the atom by tunnel ionization, was treated non-perturabative. For overlapping pump and probe pulses, the influence of the probe pulse does not decouple from the impact of the pump pulse.

Therefore, it is not clear to which extent $\rho IDM(t)$ can be extracted from the transition absorption spectrum like for non-overlapping pulses. In order to fully capture the overall effect of pump and probe pulses, both pulses are treated non-perturbatively meaning the TDCIS equations of motion are solved for an electric field $E(t)$ = Epump(t) + Eprobe(t). Note that the probe step could also be treated perturbatively by introducing a two-time IDM, which depends on two different time arguments. In our non-perturbative approach only the one-time IDM $\rho IDM(t)$ needs to be constructed for each pump-probe configuration. From $\rho IDM(t)$ the ionic dipole moment can be calculated. By performing the trace over $\rho IDM(t)$ and not over the full N-body density matrix $\rho(t)$, we consider only dipole transitions between ionic states. Transitions between virtual orbitals can be neglected, since the XUV probe pulse interacts only weakly with the excited electron. Transitions between occupied and virtual orbitals describe stimulated emission and photo-ionization processes. Both mechanism do not lead to sharp features in $\sigma a(\omega)$ around the bound-bound transition energies. Therefore, we ignore these contributions, which lead to background signals we are not interested in. The detector, where the transient absorption spectrum is measured, does not record the atomic response but rather a damped spectrum of the form $Eprobe(L,\omega)^2 = Eprobe(o,\omega)^2$ e-LnAT $\sigma a(\omega)$, where $Eprobe(o,\omega)$ is the incoming probe electric field, L is the length of the medium, $Eprobe(L,\omega)$ is the probe electric field at the end of the medium, and nAT is the atomic number density. In this Equation Beer law is used, which assumes a homogeneous medium and that the ratio hziion $(\omega)/Eprobe(\omega)$ is independent of $Eprobe(\omega)$:= $Eprobe(o,\omega)$. Due to the finite energy resolution of the detector, the transient absorption signal has to be co-evolved with a Gaussian mask function, where the full-width-at-half-maximum (FWHM) width is given by the energy resolution of the detector. The cross section $\sigma m(\omega)$ measured at the detector can be related to the atomic cross section and is given by:

$\sigma m(\omega) = -$ 1 nAT Llne $-$ nAT L$\sigma a(\omega)$ GδE(ω), where GδE(ω) is the Gaussian with the FWHM width of δE, and the symbol stands for the frequency convolution.

Oscillating Dipole Model

Here we are going to develop general expression for the transient absorption spectrum, which is based on a simple model. First, we reduce the description of the ion to a two-level system. The ground state |g> can only be accessed by the pump pulse via tunnel ionization and the excited state |e> can only be accessed by the probe pulse via resonant excitation out of |g>. The probe pulse, which may be approximated by a delta pulse, i.e., $Eprobe(t;\tau) = Eo\ \delta(t-\tau)$, creates a coherent Superposition:

$|\Psi(t > \tau)> = a_0|g> + a_1 e^{-i(\omega_0 - i\Gamma/2)(t-\tau)}|e>,$

where ω_0 is the positive energy difference between the two states, $1/\Gamma$ is the lifetime of the excited state, and $a_1 = -iE_0 a_0 z|g>$ results from the excitation by the probe pulse.

This superposition leads to an oscillating Dipole: $z>_{ion} (t > \tau)$

$=h\Psi(t)|z|\Psi(t)>$

$=-E_0|z|g>[^2 |a_0|^2 \times \sin[\omega_0 (t-\tau)]e^{-\Gamma_2 (t-\tau)}$

We see that for a simple two-level system the cross section is purely Lorentzian and directly proportional to the ground state population $|a_0|^2$ at the time of the probe step. Adiabatic energy shifts in the ionic states during the intense NIR pulse result in a phase shift in the oscillating ionic dipole, i.e., $z>_{ion} \sin[\omega_0(t \propto -\tau) + \phi(\tau)]$.

Here, we assume the ionic state and the dipole oscillation live for a long time after the NIR pulse is over such that the entire dipole dynamics can be approximated by a phase-shifted oscillation. The phase shift $\phi(\tau)$ has a dramatic influence on the shape of the transition line. Note that the phase shift $\phi(\tau)$ affects only the shape of the transition but not the strength z_0. In the case $\phi = \pi/2$, the cross section shows a dispersive behavior and has equally negative and positive regions that lie symmetrically around the field-free transition energy $\omega_0(0)$. For all other phases, the cross section is a sum of these two scenarios and becomes asymmetric around ω_0. A phase shift by π changes the sign of the cross section. For $-\pi/2 \leq \phi \leq \pi/2$, the system shows an absorbing behavior whereas for $\pi/2 \leq \phi \leq 3/2\pi$ the system is rather emitting. Similarly to the dressing of the ionic states, the influence of the excited electron on the

143

ion via the residual Coulomb interaction and via the pump field can lead to additional phase shifts in the oscillating dipole. Furthermore, corrections to the transition strength zo can occur, which may cause zo to be no longer directly proportional to the instantaneous hole population. In order to capture these effects, we parametrize in addition to the phase $\phi(\tau)$ also the transition strengths zo $zo(\tau)$.

Mechanisms leading to the Phase Shift

The dressing of the ion can induce a phase shift in the ionic dipole. In the following, we discuss in the language of TDCIS how the dressing by the field and the coupling of the excited electron to the ionic subsystem can influence the phases $\phi T(\tau)$. First, we analyze the scenario where the time evolutions of the excited electron and the ionic states are decoupled. The resulting EOM can be written as Hamiltonians of the two Subsystems which affects only the excited electron, and the ionic states. This is exactly the field-driven dressing of the ionic system. After the pulse is over [$E(t) = o$], the phase ϕion T (t,τ) becomes independent of t and depends only on the probe time τ, i.e., ϕion T (t,τ) ϕion T (τ).

Additional phase shifts similar to ϕion T (τ) can also occur due to the coupling between the ion and the excited electron. There exist two kinds of mechanism that can couple these two subsystems: (1) the residual Coulomb interaction (2) the field driven mixing of the excited N-electron states with the neutral ground state. To distinguish the phase shifts induced by the two different mechanisms, we introduce ϕresidual T (τ) and φground T (τ). The phase shift due to the residual Coulomb interaction is denoted by φresidual T (τ), and φground T (τ) denotes the phase shift due to the field driven mixing to the neutral ground state.

Phase space and quantum marginals

The Wigner distribution function In 1932 Wigner introduced a phase-space distribution function

$$\omega(q, p) = 1/(2\pi h \int \omega^*(q+\xi/2) \, e^{ip\xi/h} \, \omega(q - \xi/2)d\xi \, ,$$

(1

that fulfills the fundamental requirement for a joint probability distribution in phase space, i.e. when integrated over q or p, it gives marginal probabilities:

$\omega(q)|^2$ and $1/(2\pi h [\omega(p)|^2$, where $\omega(p) = \int dq \, e^{-ipq/h} \omega(q)$, is the Fourier transform of $\omega(q)$. As shown by Wigner, the function given by Eq. (1) is in general not positive. However, under simple and reasonable physical assumptions it is unique. Thus, when one demands that: (i) a phase space distribution P(q, p) is real, (ii) bilinear in ω, (iii) gives the correct marginals, and (iv) its dynamical evolution reproduces the Liouville equation in the classical limit, then the distribution P(q,p) is unique and is the Wigner function. After Wigner work many different phase space distribution functions have been introduced and investigated. There is a very rich literature devoted to applications of the Wigner function and other various quasi-distributions in quantum optics. The simplest example of a phase space distribution which does not satisfy the four assumptions leading to the Wigner

function, but reproduces the right marginal in an explicit way, is the distribution function:

$P(q,p) = 1/2\pi h \ |\omega(q)|^2 |\omega(p)|^2$ (2

Clearly, this distribution is not bi-linear in Ψ and as a result does not satisfy the requirements of the Wigner uniqueness theorem. Another distribution function similar to that given by Eq.2 can be guessed taking formally a square-root of this expression. As a result, up to an arbitrary phase, we have

$P(q, p) \sim \omega(q) \ \omega(p)$ (3

Note that this distribution function contains no information about the phase of the wave function. A simple insertion of an additional phase factor $\phi(q,p)$ to the wave functions leads to the expression:

$P(q,p) \sim \omega(q) e i\phi(q,p) \ \omega^*(p)$ (4

that defines a class of bi-linear but complex distribution functions. An example of such a distribution has been proposed just one year after Wigner introduced his famous distribution function.

Rydberg Atoms

In order to probe the photons, we send across the cavity a special kind of Rydberg atoms, called "circular," in which the outer electron orbits on a circle of large diameter, about a thousand times bigger than an ordinary ground state atom. These excited Rubidium atoms are prepared with lasers and radio-frequency excitation, using a modified version of a procedure invented by Daniel Kleppner and Randy Hulet at MIT in 1983. According to quantum theory, the orbiting Rydberg electron is also a wave, which has a de Broglie wavelength, and the condition of a stable orbit is that there is an integer number of these wavelengths along the circumference. This number, called the principal quantum number of the Rydberg atom, is equal to 51 or 50 in our experiments (these Rydberg states are called e and g respectively in the following). The advantage of these circular Rydberg states of maximal angular momentum over the states of small angular momentum employed in our earlier experiments is their very long natural life-time, on the order of 30 milliseconds for the states with principal quantum number 50 or 51. This life-time, of the same order of magnitude as the photon life-time in our cavity, allows us to neglect in first approximation the atomic decay processes during the interaction time between the atoms and the cavity field.

In the e and g Rydberg states, the circulating de Broglie wave has a uniform amplitude, resulting in an electron charge density centered at the atomic nucleus, yielding a zero electric atomic dipole. In order to prepare an electric dipole, a pulse of resonant microwave can be applied to the atom, bringing it in a superposition of the two adjacent e and g states, with respectively 51 and 50 nodes in their wave function. This superposition of states can be referred to as a "Schrödinger cat" because it implies an atom at the same time in two levels, reminding us of the famous cat that Schrödinger imagined suspended between life and death. A better name should be a "Schrödinger kitten," because it is made of a single atom and thus very small.

The two de Broglie waves making up this "kitten" interfere constructively at one end of the orbit and destructively at the other end, resulting in a net electric dipole, rotating in the orbital plane at 51 GHz. This dipole behaves as a rotating antenna extremely sensitive to microwave radiation. It can also be described as the rotating hand of a clock, ticking at 51 GHz. When microwave radiation, non-resonant with the transition between the two states e and g, impinges on the atom, it cannot absorb it and hence the photons remain intact, ensuring the non-demolition character essential to our experiments. However, the effect of this non-resonant light is to shift the atomic energy levels slightly and hence alter the rotating frequency of the atomic dipole, our clock hand. This light shift effect had been discovered in 1961 by Claude Cohen-Tannoudji in his seminal optical pumping studies. Light shifts are proportional to the field energy, i.e. to the photon number.

Being inversely proportional to the atom-cavity field detuning, they can be maximized by tuning the cavity close enough to resonance but far enough to avoid any photon absorption or emission process. In the case of Rydberg atoms, the effect is then very large, resulting in a phase shift of the atomic dipole after the atom leaves the cavity which can reach the value of 180°, the dipole jumping in two opposite directions when the photon number changes by one unit. Measuring this phase shift amounts to counting the photon number without destroying the light quanta. Let us note that these light shifts play an essential role in other atomic physics and quantum optics experiments. They are at the heart of the methods used to trap and cool atoms in laser light, which were recognized by the Nobel Prize awarded to Claude Cohen-Tannoudji, William Phillips and Steven Chu in 1997. In order to measure these shifts, we followed a proposal that we made in 1990. We built an atomic interferometer around our photon storing cavity. The atoms, prepared in the circular state e in the box O, cross the cavity C one by one before being detected by field ionization in D. Essential to the experiment, two auxiliary microwave zones R1and R2are sandwiching the cavity C. In the first one, the atoms are prepared in the state superposition of e and g, a "Schrödinger Kitten" state. This procedure amounts to starting a stopwatch, giving to the atomic dipole, i.e. to the clock hand, its initial direction. The atomic dipole then rotates as the atom crosses the cavity, until a second microwave flash, applied in R2, is used to detect the direction of the atomic dipole at cavity exit, thus measuring the phase accumulation of the clock. The combination of the two separated microwave resonators R1 and R2 is known as a Ramsey interferometer. The device had been invented in 1949 by Norman Ramsey. The method of separated

149

field pulses is now used in all atomic clocks working on a hyperfine microwave transition between two atomic levels. The excitation by the two successive pulses induces a sinusoidal variation of the transition probability when the microwave frequency is scanned around resonance. This so-called "Ramsey fringe" signal is used to lock the microwave frequency to the atomic transition. In our experiment, the Ramsey interferometer is counting photons by detecting the perturbing effect they produce on the fringes of a special atomic clock, made of microwave sensitive Rydberg atoms. If the phase shift per photon is set to 180°, the Ramsey fringes are offset by half a period when the number of photons changes by one. The interferometer is set at a fringe maximum for finding the atom in e if there is 1 photon in the cavity. The second pulse then transforms the state superposition of the atom exiting the cavity C either in state e (if there is 1 photon) or in state g (if there are 0 photons), this state being finally detected by the field ionization detector. The final atomic state g or e is thus correlated to the photon number, 0 or 1. In the detection events as atoms cross a cavity cooled at 0.8K, which, according to Planck's law, contains either a vacuum (95% of the time) or one photon (5% of the time). One clearly sees when a photon pops inside the cavity, stays for a while and then disappears. Due to noise and imperfections, the correlation between the photon number and the atomic signal is not perfect, but a simple majority test allows us to reconstruct without ambiguity the evolution of the photon number. The sudden change in the photon number is a quantum jump, a phenomenon predicted long ago by quantum theory and observed in the 1980s in trapped ions, as described in David Wineland's lecture. It is observed here for the first time for light quanta. In the event hundreds of atoms see the same photon between two quantum jumps, which demonstrates that our detection method is quantum non-destructive (QND) for the field. Note that these field oscillator jumps bear a strong similarity to the quantum jumps between the cyclotron oscillator states of a single electron, which were also monitored by a QND procedure. Our photon counting method can be extended to counting larger numbers of quanta. We start by injecting inside the cavity a small coherent field, a superposition of photon number states comprised between 0 and 7. This field is produced by scattering on the edges of the cavity mirrors a microwave pulse radiated by a classical source. This leads to the capture of a few photons which survive between the mirrors long after the source has been switched off. We then just need to send a sequence of atoms across the cavity, each carrying away a bit of information about the field. The phase shift per photon is optimally adjusted to a value such that different photon numbers correspond to well-separated atomic dipole directions at cavity exit. At the start of the experiment, we have no idea about the photon number and we assume a flat probability distribution, giving equal weights to the probability of having from 0 to 7 photons in the cavity. As successive atoms provide information, our knowledge about the field evolves until finally a single photon number is pinned down. The evolution of the inferred probability distribution is obtained by a Baysian argument :

each atom's measurement provides information about the atomic dipole direction and allows us to update our knowledge of the photon number distribution. This experiment shows, so to speak, live, the measurement-induced "wave-function collapse" which appears here as a progressive process transforming a flat histogram into a single peak. The field, initially in a superposition of different photon numbers, is projected by the mere acquisition of information into a photon number state, a so-called Fock state of well defined energy.

The procedure is random, The statistics of a large number of measurements reconstructs the photon probability distribution of the initial state. It obeys a Poisson law, as expected for a coherent state produced by a classical source of radiation. Once a photon number has been pinned down, its ensuing evolution can be observed by continuing the measurement with subsequent atoms on the same realization of the experiment. We then observe the succession of quantum jumps leading the field inexorably back to vacuum, due to photon losses in the cavity walls. A statistical analysis of these trajectories has allowed us to measure the Fock state life-times. Fluctuating randomly from one preparation to the next, the life-time of the n-photon number state is distributed according to an exponential probability law with the time constant Tc/n, where Tc is the cavity field energy damping time. The 1/n variation of this life-time is a manifestation of the increasing fragility of these non-classical states of radiation when their energy increases. They share this feature with Schrödinger cat states of light.

Field State Reconstruction:QND measurements

Repeated on many realizations of the same field, have allowed us to reconstruct its photon number distribution P(n), which, for a coherent field, is a Poisson function centered around its mean photon number value <n>. These P(n) histograms provide only partial information about the field. Describing the light intensity and its fluctuations, they are insensitive to field coherences. In general, a field state is described by a density matrix ρ, whose diagonal elements ρnnin the Fock state basis are the P(n) probabilities, and the off-diagonal ones, ρnn , describe the field coherence. expressed in terms of photon numbers, the P(n) probabilities are "1D" objects while the ρnn coherences are "2D" entities.

Reconstructing coherences from the measurement of photon number probabilities, i.e. "going from 1D to 2D" in the representation of the field, is analogous to going from 2D to 3D in photography. The photon number distribution of a field state is indeed like a flat photo, obtained by recording the light intensity that the object has scattered into the lens of a camera. To add an extra dimension and achieve a full reconstruction, one must realize a hologram by adding phase information to the one provided by the intensity recording. In photography, this is achieved by interfering the scattered light with a reference beam—a small fraction reflected off the main laser beam illuminating the object. The interference pattern recorded on the hologram is a Fourier transform of the object. When illuminated by a laser beam similar to the one which has produced it, the hologram reproduces the appearance of the object by inverse Fourier transformation. Similarly, the full "2D" ρnn information contained in the quantum state of a field can be reconstructed by mixing this field with reference fields of various phases and amplitudes and by reconstructing the photon number distributions of these interfering fields. This procedure is called quantum tomography. In our cavity QED experiments, the Rydberg atom Ramsey interferometer is used to perform these state reconstructions. Identical copies of the field are prepared, then admixed with reference coherent fields produced by a classical source.

QND photon counting of the resulting "mixed fields" are then performed. From the data accumulated in many realizations with reference fields of different phases and amplitude, enough information is collected to reconstruct ρ. To represent the field state, it is convenient to choose, instead of ρ, an alternative description. The field state is formally equivalent to the state of a mechanical oscillator evolving in a parabolic potential. Its state is represented by a real Wigner function taking its values in the oscillator phase space (the coordinates in this space being the position x and momentum p of the fictitious oscillator, corresponding to "field quadratures").

This function that generalizes for the quantum field the classical concept of probability distribution in phase space, contains the same information as ρ, to which it is related by Fourier transformation. To keep the holographic analogy, the Wigner function is to the density matrix what the hologram is to the direct image of an object. Its interfering patterns directly reveal the main features of the quantum field.

Schrödinger Cat States of light and decoherence Studies

By describing how we count and manipulate photons in a cavity, I have so far emphasized the 'particle aspect' of light. As was recalled above, however, light is also a wave. Which of the particle or the wave aspect manifests itself depends upon the kind of experiment, which is performed on the field. Let us describe now experiments in which the wave features of the field stored in the cavity is essential. This will lead us to the description of photonic Schrödinger cats and to decoherence experiments. At this stage, it is appropriate to recall Schrödinger's thought experiment. The Austrian physicist has imagined that a large system, a cat for instance, was coupled to a single atom, initially prepared in an excited state spontaneously decaying into a ground state by emitting a photon or a radioactive particle. This emission triggered a lethal device, killing the cat. After half the life-time of the excited state, the atom has evolved into a superposition of two states, one of which would be associated with the dead cat and the other with the live cat.

At this point, the atom and the cat would be entangled and the cat suspended between life and death. In our version of this experiment, we have a single atom in a superposition of two states and this atom controls the fate of a coherent field containing several photons our Schrödinger cat. A similar proposal for the preparation of Schrödinger cat states of light in the optical domain had been made earlier in another context.

Our method again employs the Ramsey interferometer. It starts with the preparation of a coherent field in the cavity, whose Wigner function is a Gaussian. A single non-resonant atom is then prepared in a coherent superposition of two states, an atomic Schrödinger "kitten," as I have already called it. This atom crosses the cavity and its two components shift the phase of the field in different directions by a simple dispersive index effect.

Here again, we take advantage of the huge coupling of Rydberg atoms to microwaves, which makes a single atom index large enough to have a macroscopic effect on the field phase. At the cavity exit, the atom and the field are entangled, each atomic state being correlated to a field state with a different phase (the phase difference being close to 135° in the experimental realization described below). We can consider that the field, with its small arrow in the complex plane, is a meter used to measure the atom's energy. After the atom has been exposed to the second Ramsey pulse and detected, there is no way to know in which state the atom crossed the cavity and the field collapses into a Schrödinger cat superposition. In other words, the atomic Schrödinger kitten has produced a photonic Schrödinger cat, which contains several photons on average. By sending subsequent atoms across the cavity and achieving a tomographic field reconstruction with QND photon counting, we have been able in 2008 to reconstruct completely the Schrödinger cat state Wigner function. Theory shows that the coupling of the field to the environment very quickly washes out the quantum coherence of the cat, leading it into a mundane statistical mixture of states. W. Zurek has played an important role in elucidating the role of the environment in this process, which occurs faster and faster as the "size" of the cat, measured by the square of the distance of its components in phase space, is increased. For a given phase difference between the Gaussian components, this size is proportional to the cat's mean photon number. We have studied this decoherence phenomenon by reconstructing the field Wigner function at various times. Within a time much shorter than the energy damping time of 130 ms, the interfering features of the cat state are indeed suppressed, leaving the Wigner function as a sum of two quasi-Gaussian peaks. We have checked that decoherence occurs at a rate proportional to the size of the cat. It is important to stress that these Schrödinger cat state recordings, as all field state reconstruction, are obtained from analyzing many realizations of the experiment and performing complex statistical analysis of the data. Acquiring knowledge about a quantum state always requires such a statistical procedure and these experiments rely on the fact that we can prepare an arbitrary number of copies of the state to be reconstructed and follow the subsequent evolution of all these copies. An earlier version of this experiment had been performed in 1996, with a cavity having a much shorter damping time, in the hundred microsecond range. Since it was not possible to send a sequence of measuring atoms across the cavity before its field had decayed, the experiment relied on the information provided by a single probe atom following the atom which had prepared the cat state. Instead of reconstructing the whole Wigner function, we used this single probe atom to get information about the Wigner function at phase space origin, where its value is very sensitive to the cat's coherence. Comparing the detection signal of the first atom which prepared the cat and that of the second, which probed its coherence, provided a two-atom correlation signal whose decay as a function of the delay between the two atoms measured the loss of quantum coherence of the cat versus time; There might be a shortening of the decoherence time as the separation is increased, We have also prepared Schrödinger cat states of radiation by resonant atomfield interaction. Letting a coherent field evolve under its coupling with a Rydberg atom at resonance turns the atomfield system, after some time, into an entangled atom-field state superposition involving two coherent fields with opposite phases. The two components of this cat merge together at a later time. This effect of field phase splitting and recombination is related to the collapse and revival of the Rabi oscillation phenomenon. The Schrödinger cat experiments in Cavity QED illustrate the fragility of quantum coherences in systems made of increasing number of particles. They give us a glimpse at the boundary between the quantum world, where state superpositions are ubiquitous, and the classical one, where systems behave in a mundane classical way. A detailed study of the cavity QED Schrödinger cats along with a review of proposals for the generation of various Schrödinger cats in quantum optics can be found in the book "Exploring the quantum: atoms, cavities and photons".

Serge Haroche

How to measure directly the Wigner function of the field in Cavity QED

1. Translate field in phase space by -1
2. Send a non-resonant atom with a π-phase shift per photon across Ramsey interferometer. If atom exits in level e (g), parity is even (+1) (resp.odd=-1).
3. Repeat with a large number of atoms and average between +1 (even) and -1 (odd) to find out the mean parity. This yields W(1)
4. Repeat procedure for different α values to reconstruct W and thus ρ.

   In practice, it requires to be able to realize a π-phase shift per photon for arbitrary fields, which might not be possible due to non-linear atom-field coupling. The Ramsey interferometer measures in fact a « parity operator » from which the state can be reconstructed using estimation

   procedures.

   Serge Haroche

EDM Atomic Parity Non-conservation Experiments

Physicists used to believe that parity is conserved in all interactions, and that natural laws do not distinguish between the left and the right. The discovery in 1956 that parity is not conserved had an immediate and profound influence on nuclear and elementary particle physics. However, there was no such sudden impact on atomic physics because the force that distinguishes between the left and the right, the weak interaction, is dwarfed in comparison with the electric and magnetic forces in atoms. There is no atomic process, as in nuclear β-decay, in which the weak interaction is uniquely manifested. Only recently, over two decades after the original parity revolution, has parity non-conservation PNC been observed in the electronic structure of atoms. Atomic PNC is caused by an interesting form of the weak interaction not known in 1956, the neutral current form in which the interacting particles (electrons and nucleons in the case of atoms) do not change charge. These weak neutral currents are a central prediction of the unified theory of weak and electromagnetic forces developed by Glashow, Weinberg, Salam and others in the 1960's. One of the important tests of this theory has been to find the expected effects of the neutral current interaction in atoms. Now that these effects have been observed, atomic PNC measurements should continue to play a useful role in working toward a complete understanding of weak neutral currents. An important symmetry that characterizes the unified theory in its present form is time reversal invariance (T-symmetry). Likewise, the observed atomic PNC effects, such as optical rotation, possess T-symmetry. A different type of PNC effect that violates T-symmetry is a permanent atomic electric dipole moment. Such EDM has been sought in neutrons and in atoms and molecules. EDM confirm the long suspected existence of T-violating forces between elementary particles. As early as 1959 Zeldovich estimated the size of optical rotation due to a neutral current form of the weak interaction that might be expected in a gas of ground state atomic hydrogen and concluded that this particular effect would be far too small to observe. Later Curtis-Michel analyzed some possible experiments with excited states of hydrogen which would take advantage of the close proximity of states of opposite parity, but at that time the experiments still seemed to be very difficult. Two serious attempts were made in the 1960 to observe PNC in atomic systems by looking for circular polarization associated with magnetic-dipole transitions. Upper limits were set for molecular oxygen and atomic lead, but neither experiment approached the sensitivity required to observe weak interaction effects. Meanwhile, extraordinary advances in the field theory of weak interactions during the 1960 strengthened the theoretical basis for weak neutral currents. Renormalizable gauge theories of the interactions among elementary particles led to a unified picture of the weak and electromagnetic forces. The simplest of these theories was that of Weinberg and Salam, often called the standard model" of electroweak interactions, in which the weak interaction between any pair of fermions is mediated by the exchange of massive vector bosons, two charged ones, W-+', and one neutral one, Z $^-$. Only one free parameter is introduced, the mixing angle between the bare Z $^-$ and bare photon, called the Weinberg angle, Θw. The charged bosons mediate the familiar form of weak interaction known in the earlier language of current-current interaction as the charged or perhaps better, charge-changing current process. The best known example is ordinary β-decay, in which a neutron is converted into a proton while an antineutrino and an electron are created, i.e. n p+e+v. The equivalent scattering process is n+v p+e. In both cases the neutron-proton current acts through a W-+ with the neutrino-electron current. The weak neutral currents predicted by the Weinberg-Salam theory are mediated by the neutral Z −. In the case of atoms, this neutral current allows the direct first-order weak interaction of the atomic electrons with the nucleons and with each other without changing the atom's identity. For example, e+p e+p, e+n e+n, and also e+e e+e. In these cases, the electron-electron current acts through the Z − with either the protonproton, neutron-neutron, or electron-electron current. In 1973 high energy neutrinc+nucleon scattering experiments demonstrated the existence of such weak neutral currents in nature. A variety of experiments have subsequently shown neutral current phenomena, all consistent with predictions of the Weinberg-Salam theory. The PNC neutral current interaction between electrons and nucleons has been established both by measurements of high energy inelastic electron scattering from protons and deuterons, and by the atomic PNC experiments. In addition, two groups using the CERN colliding beam facility have observed both the W-+ and the Z − resonances directly in proton-antiproton collisions. Recent atomic parity non-conservation PNC experiments Atomic PNC experiments under consideration in many laboratories received vital encouragement in 1974 when Bouchiat and Bouchiat pointed out that there should be considerable enhancement of neutral current effects in heavy atoms. They showed that PNC effects should increase with atomic number Z roughly as Z_3, and demonstrated that the heavy atoms Cs and Tl should exhibit optical helicity more than 6 orders of magnitude greater than ground-state hydrogen. This work provided impetus to world-wide experimental efforts with Cs, Bi and Tl.

Later, after attention was drawn once again to experimental possibilities with metastable atomic hydrogen, a number of experiments to measure PNC in hydrogen also began. After several years of intensive effort the original goal of detecting and studying weak neutral currents in atomic physics has been realized. Many experiments with heavy atoms have attained enough sensitivity to see PNC effects of the size expected. Over the same period the complex atomic calculations with heavy atoms have been refined considerably. Experimental results agree in sign, and most agree in approximate magnitude, with the predictions of the most recent atomic calculations using the Weinberg-Salam theory. However, there remain some discrepancies among the experimental results that need to be clarified, and we still await the first results from the important atomic hydrogen experiments. Improvements in experiments are being implemented which should set more stringent limits on alternative gauge theories, detect certain neutral current couplings not yet observed, and likely test the predicted contributions of Z $^-$ and W-+ radiative corrections. There have also been recent limits set by experiments searching for a permanent atomic or molecular EDM and new atomic EDM experiments have begun, with the goal of finding another manifestation of the CP violation observed long ago in K $^-$

decay.

Observable effects : PNC-induced electric dipole moments
The inversion symmetry of a physical system and in a similar
fashion the intrinsic symmetry of an elementary particle can be
described with the use of a parity operator; P. If the state
function of the physical system is, ω, the operator P inverts the
spatial coordinates of the function, so that $P\omega(r)=\omega(-r)$. If
$P\omega(r)=+-\omega(r)$, then we say that the function ω has well defined
parity (even parity if +,odd parity if -). We also say that parity
is conserved by an interaction if the Hamiltonian, H, for that
interaction commutes with the parity operator, or equivalently,
if H is invariant under coordinate inversion. The measured
quantity in atomic PNC experiments which search for the
inversion asymmetry is an electric dipole moment, induced in
an atom by a force which violates parity conservation. When
this force violates time-reversal symmetry as well, dipole
moment can be a permanent electric dipole moment EDM,
which causes an energy shift of the atom in an external electric
field. If instead the force obeys time-reversal symmetry, the
Dipole is restricted to being a transition dipole moment which
is observable through its interference with some other atomic
moment in radiative transitions between atomic states. We
begin with a simplified discussion. First suppose there is a
permanent electric dipole fixed parallel to the total atomic
angular momentum F in a non-degenerate stationary state of
an atom. In an external field E the negative energy shift of the
Dipole, E, proportional to F.E, clearly violates both P- and T-

164

symmetry since F F, E -E under P, while F -F, E E under T.

Thus an observation of a permanent EDM parallel to an atomic spin would be clear evidence of an interaction in the atom which violates time-reversal invariance. Next suppose that T-symmetry is not violated. Since we will observe no permanent electric dipole moment of the atom, how can we observe a PNC effect?

We should look instead for oscillating dipole moments associated with transitions between atomic states. For example, suppose a magnetic dipole transition takes place with an oscillating atomic magnetic moment M. The PNC force within the atom can induce the Dipole with a component parallel to M, but T-symmetry requires these two oscillating moments to differ in phase by $\pi/2$; otherwise the T-odd quantity of the Dipole moment M would not average to zero. This phase difference in El and Ml radiation causes circular polarization, an observable PNC effect in the case of time-reversal symmetry.

Conservation of Baryon Number

Baryons are hadrons, i.e. composite particles made of quarks composed of any three quarks. Baryon number is conserved in a reaction. You should count each baryon as +1 and each antibaryon as -1. Non-baryons have a baryon number of 0. p- + p+ n° + p- + p+ This is an observed event that conserves both electric charge and baryon number. p+ p+ + p° → This conserves charge but not the baryon number.

Conservation of Lepton Number

165

The electron is the best known lepton. The tau and the muon are the other two charged leptons. Each neutrino is associated with one of the charged leptons. Lepton number is also conserved in reactions. Again, leptons have lepton number of +1, antileptons have -1, and non-leptons have 0. e+ + e- p+ + p→ This is an observed event that conserves both electric charge and baryon number. p- e- + g → Charge is conserved, but lepton number is not. There are no leptons on the left, but there is one on the right.

Quantum Jump

Measurements are described with diverse concepts in quantum physics such as; wave functions/probability amplitudes, evolving unitary and deterministic/preserving information, according to the linear Schrödinger equation, superposition of states, i.e., linear combinations of wave functions with complex coefficients that carry phase information and produce interference effects/the principle of superposition, quantum jumps between states accompanied by the "collapse" of the wave function that can destroy or create information, probabilities of collapses and jumps given by the square of the absolute value of the wave function for a given state, values for possible measurements given by the eigenvalues associated with the eigenstates of the combined measuring apparatus and measured system, the Heisenberg indeterminacy principle. The original problem, said to be a consequence of Niels Bohr's "Copenhagen interpretation" of quantum mechanics, was to explain how our measuring instruments, which are usually macroscopic objects and treatable with classical physics, can give us information about the microscopic world of atoms and subatomic particles like electrons and photons.

Bohr's idea of "complementarity" insisted that a specific experiment could reveal only partial information, for example, a particle's position. "Exhaustive" information requires complementary experiments, for example to determine a particle's momentum.

Some define the problem of measurement simply as the logical contradiction between two laws describing the motion of quantum systems; the unitary, continuous, and deterministic time evolution of the Schrödinger equation versus the non-unitary, discontinuous, and indeterministic collapse of the wave function. Von Neumann saw a problem with two distinct procedures. The mathematical formalism of quantum mechanics provides no way to predict when the wave function stops evolving in a unitary fashion and collapses. However, we can say this occurs when the microscopic system interacts with a measuring apparatus. Others define the measurement problem as the failure to observe macroscopic superpositions. Decoherence theorists, e.g., H. Dieter Zeh and Wojciech Zurek, who use various non-standard interpretations of quantum mechanics that deny the projection postulate - quantum jumps - and even the existence of particles, define the measurement problem as the failure to observe superpositions such as Schrödinger's Cat.

Unitary time evolution of the wave function according to the Schrödinger wave equation should produce such macroscopic superpositions, they claim. Information physics treats a measuring apparatus in quantum mechanical terms by describing parts of it as in a metastable state like the excited states of an atom, the critically poised electrical potential energy in the discharge tube of a Geiger counter, or the supersaturated water and alcohol molecules of a Wilson cloud chamber; namely, the pi-bond orbital rotation from cis- to trans- in the light-sensitive retinal molecule is an example of a critically poised apparatus.

According to the correspondence principle, all the laws of quantum mechanics asymptotically approach the laws of classical physics in the limit of large quantum numbers and large numbers of particles; Quantum mechanics can be used to describe large macroscopic systems.

Does this mean that the positions and momenta of macroscopic objects are uncertain? Yes, it does, though the uncertainty becomes vanishingly small for large objects, it is not zero. Niels Bohr used the uncertainty of macroscopic objects to defeat Albert Einstein's several objections to quantum mechanics at the 1927 Solvay conferences.

But Bohr and Heisenberg also insisted that a measuring apparatus must be a regarded as a purely classical system. They can't have it both ways. Can the macroscopic apparatus also be treated by quantum mechanics or not? Can it be described by the Schrödinger equation? Can it be regarded as in a superposition of states?

The most famous examples of macroscopic superposition are perhaps Schrödinger's Cat, which is claimed to be in a superposition of live and dead cats, and the Einstein-Podolsky-Rosen experiment, in which entangled electrons or photons are in a superposition of two-particle states that collapse over macroscopic distances to exhibit properties at speeds faster than the speed of light. These treatments of macroscopic systems with quantum mechanics were intended to expose inconsistencies and incompleteness in quantum theory. The critics hoped to restore determinism and "local reality" to physics. They resulted in some strange and extremely popular "mysteries" about "quantum reality," such as the "many-worlds" interpretation, "hidden variables," and signaling faster than the speed of light.

We develop a quantum-mechanical treatment of macroscopic systems, esp. a measuring apparatus, to show how it can create new information. If the apparatus were describable only by classical deterministic laws, no new information could come into existence. The apparatus need only be adequately determined, that is to say, "classical" to a sufficient degree of accuracy.

Everything said so far indicates the sensible which is the performance of quantum computing to the correct measurement of the quantum states.

On the other hand, a new technology allows us to avoid the problem of quantum measurement, However, this technology allows work exclusively with Computational Basis States (CBS), i.e., pure and orthogonal quantum base states.

Therefore, a new method of quantum measurement in the case of generic qubits becomes imperative and more accurate than the methods currently in use. Thus, here we present a novel proposal to recover quantum state to the output of a quantum algorithm after its measurement via a modified Kalman's Filter, and Recursive Least Squares (RLS) filter, too. Here we are going to present the following topics:

- Wave function collapse
- Quantum Measurement Problems
- Before and after measurement
- Types of measurement and state reconstruction

Wave function collapse

In quantum mechanics, wave function collapse is the phenomenon in which a wave function -initially in a superposition of several eigenstates- appears to reduce to a single eigenstate after interaction with a measuring apparatus. It is the essence of measurement in quantum mechanics, and connects the wave function with classical observables like position and momentum. Collapse is one of two procedures by which quantum systems evolve in time; the other is continuous evolution via the Schrödinger equation; However in this role, collapse is merely a black box for irreversible interaction with a classical environment. Calculations of quantum decoherence predict apparent wave function collapse when a superposition forms between the quantum system's states and the environment's states. Significantly, the combined wave function of the system and environment continue to obey the Schrödinger equation.

When the Copenhagen interpretation was first expressed, Niels

Bohr postulated wave function collapse to cut the quantum world from the classical. This tactical move allowed quantum theory to develop without distractions from interpretational worries. Nevertheless it was debated, for if collapse were a fundamental physical phenomenon, rather than just the epi-phenomenon of some other process, it would mean nature were fundamentally stochastic.

This issue remained until quantum decoherence entered mainstream opinion after its reformulation in the 1980s. Decoherence explains the perception of wave function collapse in terms of interacting large- and small-scale quantum systems, and is commonly taught at the graduate level; the Cohen-Tannoudji textbook. The quantum filtering approach and the introduction of quantum causality non-demolition principle allows for a classical-environment derivation of wave function collapse from the stochastic Schrödinger equation.

Quantum Measurement Problems

The measurement problem in quantum mechanics is the unresolved problem of how or if wave function collapse occurs. The inability to observe this process directly has given rise to different interpretations of quantum mechanics, and poses a key set of questions that each interpretation must answer. The wave function in quantum mechanics evolves according to the Schrödinger equation as a linear superposition of different states, but actual measurements always find the physical system in a definite state.

Any future evolution is based on the state the system was

discovered to be in when the measurement was made, meaning that the measurement "did something" to the process under examination. Whatever that "something" may be does not appear to be explained by the basic theory.

To express matters differently, the Schrödinger wave equation determines the wave function at any later time. If observers and their measuring apparatus are themselves described by a deterministic wave function, why can we not predict precise results for measurements, but only probabilities? As a general question: How can one establish a correspondence between quantum and classical reality?

Before and after measurement

In quantum mechanics, measurement is a non-trivial and highly counter-intuitive process. First, because measurement outcomes are inherently probabilistic, i.e. irrespective of the carefulness in the preparation of a measurement procedure, the possible outcomes of such measurement will be distributed according to a certain probability distribution; Secondly, once a measurement has been performed, a quantum system in unavoidably altered due to the interaction with the measurement apparatus. Consequently, for an arbitrary quantum system, pre-measure and post-measure quantum states are different in general.

Measurement Postulate

Quantum measurements are described by a set of measurement operators {Mm, index m labels the different measurement outcomes, which act on the state space of the system being measured. Measurement outcomes correspond to values of *observables*, such as position, energy and momentum, which are Hermitian operators corresponding to physically measurable quantities.

Let [ω) be the state of the quantum system immediately before the measurement. Then, the probability that result *m* occurs is given by:

$p(m) = (\omega] M' M [\omega)$ (1)

Operators M*m* *should* satisfy the completeness relation, i.e., Σ $M' M = I$ because that guarantees that probabilities will sum to one: $\Sigma (\omega] M' M [\omega) = \Sigma p(m)$.

Let's Assume we have a polarized photon with associated polarization orientations 'horizontal' and 'vertical'. The horizontal polarization direction is denoted by [0) and the vertical polarization direction is denoted by [1) . Thus, an arbitrary initial state for our photon can be described by the quantum state [ω) =a[0) + b[1) , where a and b are complex numbers constrained by the normalization condition $a^2+b^2=1$ and { [0) , [1) } is the computational basis spanning H^2.

Now, we construct two measurement operators M0=[0)(0] and M1=[1)(1] and two measurement outcomes a0,a1. Then, the full observable used for measurement in this experiment is M=a0[0)(0]+ a1[1)(1].

According to Postulate, the probabilities of obtaining outcome a0 or outcome a1 are given by

p(a0)=a^2 and p(a1)=b^2 . Corresponding post-measurement quantum states are as follows: if outcome = a0 then

$[\omega)pm=[0)$; if outcome= a1 then $[\omega)pm=[1)$

Types of measurement and state reconstruction
From a practical point of view, inside context of quantum image processing, the problem is reduced to the following: suppose we develop a quantum algorithm for filtering classic images. A first problem would be, how to introduce a classical noisy image within the quantum computer? That is to say, design of the interfaces (classical-to-quantum, and quantum-to classical). The second problem would be, how to measure the results of a quantum filtering algorithm, and to take the result of that filtering process and carry out to the classical world, in other words, the recovery of the classical version of the filtered image into its original space, i.e., the classic world where it was generated. It is obvious that an absolutely accurate technique of measurement is needed.
In the last decade there have been several efforts to remedy this situation including:
- Weak measurement
- Restoring the quantum state
- Quantum state tomography
Weak measurements differ from normal (sometimes called "strong" or "von Neumann") measurements in two ways:

1. If $[\omega)$pm has discrete spectrum, a strong measurement when the system is in state $[\omega)$ yields an eigenvalue of $[\omega)$pm; if the measurement is repeated many times; starting each time with the system in state $[\omega)$, one obtains a sequence of eigenvalues

of [ω)pm that when averaged yield an
approximation to (ω]ω pm[ω), the expectation of [ω)pm in the
state [ω) .

By contrast, a *weak measurement* only yields a sequence of
numbers which average to (ω]ω pm[ω) . For instance, a strong
measurement of the spin of a spin-1/2 particle must yield spin
1/2 or -1/2.

2. A strong measurement changes ("projects") an initial pure
state [ω) to an eigenvector of
[ω)pm . (The particular eigenvector obtained cannot be
predicted, though its probability is determined.) This
substantially changes the state y unless y happened to be close
to that eigenvector, However, a weak measurement does not.
Weak measurements are often implemented by coupling the
original system W to be measured with an auxiliary quantum
"meter system" *M*. The meter along a scale, though in practice
various microscopic quantum systems are used. The composite
system is mathematically represented as the tensor product of
ω with *M*, denoted ω⊗*M* . A "product" state in this tensor
product is typically denoted [ω)[M).

Restoring the Quantum state
Restoring the quantum state is an effort to recover the original
state [ω) from the alleged invertibility of measurement
operator through the matrix that represents, that is to say *M*.
Parrott work is presented in opposition to the technique of
weak measurement in general. Based on Stochastic Procedures

and Adaptive Filtering the single matrix inversion in the procedure of estimation or identification does not restore the state of a system hidden behind such matrix.

This is due to the need to model correctly state and measurement noises and the appropriate architecture of the estimator for the correct system state recovery from the observables. This deficiency explains why Wiener filter was completely replaced by the Kalman's filter in the presence of said noise. Therefore, this technique is as weak as that at which it opposes.

Quantum state tomography is the process of reconstructing the quantum state (density matrix) for a source of quantum systems by measurements on the systems coming from the source. Being the density matrix for pure or mixed states,

$\rho = \sum p(m) \, [\omega m)(\omega m]$

The source may be any device or system which prepares quantum states either consistently into quantum pure states or otherwise into general mixed states. To be able to uniquely identify the state, the measurements must be in tomographic terms complete. That is, the measured operators should form an operator basis on the Hilbert space of the system, providing all the information about the state. Such a set of observations is sometimes called a quorum. In quantum process tomography on the other hand, known quantum states are used to probe a quantum process to find out how the process can be described. Similarly, quantum measurement tomography works to find out what measurement is being performed. The general principle behind quantum state tomography is that by repeatedly performing many different measurements on quantum systems described by identical density matrices, frequency counts can be used to infer probabilities, and these probabilities are combined with Born's rule to determine a density matrix which fits the best with the observations.

A proposal for the optimal estimation of states in Quantum Information Processing

Mario Mastriani

CIT-DGIIT-ANSES, 353 Piedras St., 2nd. Floor, Room 228 (C1070AAG), Buenos Aires, Argentina.

mmastri@anses.gov.ar

Preservation of phase space volume and Liouville's theorem

The theoretical utility of the Hamiltonian is Liouville's Theorem. In Classical Mechanics, the complete state of a particle can be given by its coordinates and momenta. For example in three dimensions, there are three spatial coordinates and three conjugate momenta. If we consider a six dimensional phase space, a point in that space represents the state of a particle. A particle will follow a determined path through phase space, that is, given the particles a point in phase space, our equations of motion will yield the phase space location of the particle at a later time or earlier time. So particles follow determined paths through six dimensional phase space. Now consider a large number of particles in a beam. These particles can be described by points in phase space per particle. For large numbers of particles in a system, or if we consider a theoretical ensemble of particles, the system can be described as a density group which is a function of the position in phase space.

Two Series for Gamma Function

It does not seem to be widely recognized that the Stirling asymptotic series for $T(x)$ yields accurate amounts for small integer arguments. However, Salzer has pointed out the effectiveness of this series in approximating $T(z)$ for large amounts of z even when $R(z)$ is quite small. Though the Stirling series for ln $T(z)$ contains only odd powers of zr1, whereas the corresponding series for $T(z)$ contains all powers of z~1, nevertheless the latter provides an effective computational tool for the direct evaluation of $T(z)$, esp. by means of modern digital. For that reason, the precise rational amounts of the first coefficients of Stirling's asymptotic series for $T(z)$ have been calculated and are tabulated herein. The second series here considered would be the power series for the entire function $1/T(z)$. The first extensive calculation of the coefficients of this series appears to recalculated and corrected by Isaacson and Salzer. These emended amounts have been reproduced in Davis and in the NBS Handbook. In the course of checking these corrected amounts the present author has recalculated these coefficients and extended the approximations to 31D.

The Riemann States

Consider the quantum state

$$|\omega(o)>=|\omega os)>* \ |\omega at)>-\Sigma \omega n \ |n> * \ ce \ |e>+cg \ |$$

g)>, (1

which is the direct product of the initial oscillator and atomic state, ϕ and $|\omega at)>$, respectively. The harmonic oscillator can be represented by a single mode of a cavity field or the motion of a trapped ion. At time t-o, it is given by a superposition of Fock states $|n>$ with probability amplitudes ωn. . Likewise, the atomic state is in a superposition of the ground and the excited state, $|g)>$ and $|e>$, of a two-level atom with probability amplitudes cg and ce. Both states have to be normalized to ensure the probability interpretation which implies $\sum \omega n[^2-1$ and $ce[^2+cg[^2-1$.

The time evolution arises due to the Riemann Hamiltonian; $HR= h\omega \ln(n+1) \sigma z$; Here, ω denotes the Rabi frequency which establishes the coupling between the atom and the field mode and $\sigma z= |e><e|-|g><g|$ is the Pauli spin matrix. The Hamiltonian HR is reminiscent of the effective Hamiltonian of the Jaynes-Cummings-Paul model: $HJC=h\omega n\sigma z$, which is central to cavity QED.

With the Riemann Hamiltonian HR, we get: $|\omega(t)>=|\omega e(t)> |e>+|\omega g(t)> |g>$, for the time-evolved state. We construct states which reproduce the Dirichlet series, $\zeta(s)=\sum 1/n^\wedge s$ of the zeta function by a joint measurement that is the the overlap C(t)-$<\phi|\omega(t)>$ between the time-evolved state $| \omega(t)>$ and the entangled reference state $|\phi\rangle = \sum|\phi a\rangle \; |a\rangle$ along with the Wigner function which then provides us with the overlap of two pure states $\langle\phi|\omega\rangle^2$. It is crucial to keep in mind that this representation is considered to be applied for $\sigma>1/2$, since this fact restricts the quantum states to the same region; Therefore, we name these states Riemann states. Moreover, we analyze the behavior of their phase space representations.

The Dirichlet

When we rewrite the Dirichlet sum $\zeta(s)=\sum 1/n^s$, by using $s=\sigma+i\tau$ and shifting the summation index, it becomes evident that $\zeta(s)-\sum 1/(n+1)^\sigma$ [$e-i\tau \ln(n+1)$ holds some similarities with the quantum mechanical

time evolution, $|\omega(t)\rangle = |\omega e(t)\rangle \; |e\rangle + |\omega g(t)\rangle \; |g\rangle$ (2

Indeed, choosing ce-1 and cg-0, the overlap $C(t)$ between the time-evolved state $|\omega(t)\rangle$ and the entangled reference state $|\phi\rangle$ yields $C(t)-\sum\phi ne \; \omega n \; e-i\omega t \ln(n+1)$ (3

We can reproduce the Dirichlet representation (3, by eq. (4, From the scalar product $\langle\phi|\omega\rangle$ emerges the overlap of the state

$|\omega(t)\rangle - N(\sigma)\sum 1/(n+1)^\sigma/2$ [$e-i\omega t \ln(n+1) \; |n,e\rangle$ (4

and its initial state $|\phi\rangle$ to be applied as a reference. The normalization then gives

$N(\sigma)-[1/\sqrt{\sum 1/(n+1)^\sigma}] - 1/\sqrt{\zeta(\sigma)}$ (5

which is convergent for $\sigma>1/2$. thus with these states we can describe the zeta function in the region where the Dirichlet sum is convergent. The equation (4, clearly shows that the time-evolved state $|\omega(t)\rangle$ would remain a product state of the

180

excited state $|e>$ and the state $|\sigma,\tau>$ at the scaled time τ-ωt. Thus the overlap reduces to the scalar product of the time-evolved state $|\sigma,\tau>$ with its initial state implying that entanglement is not necessary for the description of the zeta function in the region $\sigma>1/2$, Therefore, we call $|\sigma,\tau>$ the Riemann state. It is worthwhile to mention that $|\sigma, \tau>$ constitutes the thermal phase state:

$|\omega p>=Np\sum e$-$\sigma n/2$ $|n>$; (6

Both states are coherent superposition of photon number states with real expansion coefficients which decay with n. Their similarities and differences become evident, when we compare the corresponding photon distributions: In the case of the thermal phase state $|\omega p>=Np\sum e$-$\sigma n/2$ $|n>$, we obtain: $[<n|\omega p>[^2$ which decays exponentially with the photon number n and has a max. at at n-0 that is for the vacuum state $|0>$. For the Riemann state the max. of $[<n|\sigma,0>[^2$- $[N(\sigma)[^2/(n+1)^\sigma$- $[N(\sigma)[^2$ e-$\sigma ln(n+1)$; Would be at n-0; However, the photon statistics only decays polynomial with n, which is much slower than the exponential decay. In this sense, we have replaced n by ln $(n+1)$ in eq. (6, in agreement with the construction of the Hamiltonian:

$HR=h\omega \ln(n+1)$ σz.

The Cohen distribution functions

We have already emphasized the marginal properties of a phase space distribution function. An interesting question arises about a general form of the distribution with the correct marginals. This problem has been posed and solved by Cohen

181

in 1966. The most general distribution P(q,p) with the proper marginals has the form of a double Fourier transform of a function

$A(q',p') = \int e^{-ip'\xi/h} \Psi^*(\xi'-q/2) \Psi(\xi+q/2) d\xi$,

multiplied by an arbitrary function $\Phi(q',p')$ satisfying the relations $\Phi(q',0) = \Phi(0,p')$

In the literature devoted to optical processing of classical signals, the function $A(q',p')$ is called the Ambiguity function. Therefore, the Cohen joint distribution functions labeled by functions Φ are given by the following equation

$P\Phi(q,p)=F[\Phi A[:= 1/(2\pi h)^2 \int\int e^{i(p'q-q'p)/h} \Phi(q,p) A(q,p) dp\, dq$,

where by F we have denoted a double Fourier transform. The Wigner distribution function is obtained by substituting $\Phi(q',p') = 1$; in the Cohen distribution functions formula. The distribution function is obtained for $\Phi(q',p)=\exp[-ip'q'/2h$; $\Phi(q',p')=\cos p'q'/2h$ leads to the Margenau-Hill.

The Kirkwood-Rihaczek distribution function

In 1933 Kirkwood introduced a phase space distribution which, according to his description, ". . .differs but little from the Wigner function " . The Kirkwood function is defined as follows:

$K(q,p) = 1/2\pi h \int d\xi\, \omega(q) e^{i(\xi-q)p/h} \omega^*(\xi) = 1/2\pi h\, \omega(q) e^{-ipq/h} \omega^*(p,$ (1

In 1968, this function was rediscovered by Rihaczek in the context of a signal energy distribution in time and frequency. The real part of the Kirkwood-Rihaczek distribution

$KRe(q,p) = Re[\omega(q) e - ipq/h \omega^*(p)$

is closely related to a quantum mechanical phase space distribution introduced by Margenau and Hill .

We clearly see that the Kirkwood-Rihaczek distribution has the correct marginal properties:

$\int K(q,p) dp = |\omega(q)|^2$, $\int K(q,p) dq = 1/2\pi h |\omega(p)|^2$ (2

Another condition easy to find is that the absolute square of $K(q,p)$ has the form similar to the following equation:

$|K(q,p)|^2 = 1/(2\pi h)^2 |\omega(q)|^2 |\omega(p)|^2$,

that indicates that K(q, p) is a square-integrable function. The absolute square of the function, has a simple physical interpretation. It is just proportional to the product of the probabilities in configuration and momentum representations. The dynamical free evolution of a particle with mass m of the distribution function is given by the following equation $\partial t K(q, p, t) + p/m \partial q K(q, p, t) = ih/2m \partial^2 q K(q, p, t)$ that can be also written in a form:

$K(q, p, t) = et(ih/2m \partial^2 q - p/m \partial q) K(q, p, 0)$;

We see from this formula, that the free evolution of the function is a superposition of the free Schrödinger diffusion and of a classical boost to a moving frame. The distribution function is bi-linear and has the correct marginal properties but, in contrast with the Wigner function, it is not real nor does its free evolution satisfy the classical Liouville equation. As we shall see in detail later, it is also not well-behaved under rotations of the (p,q) coordinate system, and this implies that such a function can not be measured by tomography methods. The simplicity of the definition, Eq. (1, indicates that it is relatively easy to evaluate the distribution function even for systems for which an analytical formula of the Wigner function is not known,The best example of which is a Hydrogen atom.

Non-classical Properties of Coherent States

It is demonstrated that a weak measurement of the squared quadrature observable may yield negative values for coherent states. This result cannot be reproduced by a classical theory where quadratures are stochastic c-numbers. The real part of the weak value is a conditional moment of the Margenau-Hill distribution. The non-classical term of coherent states can be associated with negative values of the Margenau-Hill distribution. A more general type of weak measurement is considered, where the pointer can be in an arbitrary state, pure or mixed. Harmonic oscillator coherent states were first investigated by Schrödinger, who was looking for classical-like states. There are several ways in which coherent states are the most classical of any pure state. They keep their shape, not spreading out as they move in the harmonic oscillator potential.

They minimize Heisenberg uncertainty relation, with equal uncertainty in both quadratures; In this way, they are the closest possible quantum mechanical representation of a point in phase space.

The term coherent state was introduced by Glauber; He demonstrated that coherent states are produced when an essentially classical current interacts with the radiation field . Aharonov demonstrated that coherent states are the only pure states that produce independent output when split in two. Zurek have demonstrated that coherent states are natural pointer states for a harmonic oscillator weakly coupled to a thermal environment. Glauber and Sudarshan demonstrated that any density operator can be expanded in terms of coherent states: $\rho = \int d^2\alpha P(\alpha) |\alpha)(\alpha|$. (1

The weight function $P(\alpha)$ is known as the P-distribution. Glauber defined non-classical states as those for which the P-distribution fails to be a probability density. More specific, non-classical states have a P-distribution which is negative or more singular than a δ function. It is the purpose of this part to demonstrate that a quantum state may be non-classical even though the P-distribution is a probability density, and that also coherent states display non-classical characteristics. In this part, we associate non-classical with the failure of the Margenau-Hill distribution to be a probability distribution. The Margenau-Hill distribution yields correct marginal distributions, just as the Wigner distribution; But in contrast to the Wigner distribution, it is negative for coherent states. We give an operational significance to conditional moments of the Margenau-Hill distribution by demonstrating that they can be observed in weak measurements. Weak measurements were proposed by Aharonov. Their suggestion was initially met with criticism, but has since been confirmed in various ways. The results reported here are related to a paper by Aharonov; which demonstrated that a weak measurement of kinetic energy of a particle in a classic forbidden region might yield negative amounts. In the original von Neumann measurement scheme, it was found that in order to distinguish different eigenvalues of the object, the pointer should be in a state with small uncertainty in the pointer position. Aharonov proposed to define weak measurements by using a pointer with a large pointer position uncertainty. Here, we abandon this condition. Instead, we assume that the interaction between the pointer and the object is sufficiently weak. Thus, the pointer can be in an arbitrary state, pure or mixed. We impose only one condition on the pointer, namely that the current density should vanish. We consider an object and a pointer described by the density operators ρs and ρa, respectively. Prior to the measurement interaction, the combined object plus pointer is assumed to be in a product state $\rho o = \rho s \otimes \rho a$. We wish to perform a weak measurement of an arbitrary object observable c. To this end, we shall assume that the interaction part of the Hamiltonian has the form: $He = \varrho\, \delta(t) c \otimes p$

Here includes two equal important parts:

    -The interaction Hamiltonian in its essence is the same, except that we have introduced an interaction strength ϱ . It is a specification of the interaction Hamiltonian proposed by von Neumann

        -During the measurement interaction, the interaction part of the Hamiltonian dominates the time evolution.

Representations In Probability/Phase Space:
We present in this part a brief review of the concept of density operator. For a pure state $|\omega\rangle$, the density operator is defined

by $\rho=|\omega><\omega|$. If instead one is uncertain about the state of the system ,and we know that there is a probability $P\psi$ for the system to be in state $<\omega|$, the density operatoris defined by

$\rho=\sum P|\omega><\omega|$. (1

The utility of the above definitions can be grasped by writing down,in terms of ρ,the average value of an observable A

$A=\sum P<\omega|A|\omega>=Tr(\rho A)$ (2

that represents a unified way of expressing the average value, valid both for a pure state and a statistical mixture. From the definition it follows immediately that $Tr\rho=\sum P\omega=1$. Also, it is easy to see that ρ is Hermitian, and therefore can be diagonal. If $|\phi i>$ are the eigenstates of ρ, then

$\rho=\sum P|\phi i><\phi i|$, $<\phi i|\phi j>=\delta ij$, that implies:

$\rho^2=\sum P^2|\phi i><\phi i|$ $Tr\rho^2=\sum P^2\leq 1$.

For a pure state, $Tr\rho^2=1$, while for a mixture $Tr\rho^2<1$. In terms of the Fock basis, $\rho=\sum \rho nm|n><m|$. Let us recall that if one has two systems (interacting or not), let's say an atom and an electromagnetic field, a basis for the combined system can be obtained by forming the tensor product of the bases corresponding to each of the two systems. The tensor product of two states $|\omega A>$ and $|\omega F>$ corresponding respectively to the systems A and F is written as $|\omega A>$ $|\omega F)$, corresponding to the density operator $\rho=\rho A\otimes \rho F$. The average of expressions involving products of operators acting on A and on F separately can be written as $(AF)=Tr(\rho A)Tr(\rho F$, The general state of the combined system will not have the form of a tensor product, but can be expressed as a linear combination of tensor product states.

187

A Review of Quantum information with real or artificial atoms and photons in cavity

The Phase Qubit

Rydberg atoms in states e and g behave as qubits, They are prepared in B in state e and cross one at a time the high-Q cavity C where they are coupled to a field mode. The atom field system evolution is ruled by the Jaynes-Cummings hamiltonian; A microwave pulse applied in R_1 prepares each atom in a superposition of e and g, After C, a second pulse applied in R_2 , maps the measurement direction of the qubit along the Oz axis of the Bloch sphere, before detection of the qubit by selective field ionization in an electric field in D. The R_1-R_2 combination constitutes a Ramsey interferometer. This set-up has been used to entangle atoms, realize quantum gates, count photons non-destructively, reconstruct non classical states of the field and demonstrate quantum feedback procedure.

Quantum Information with real or artificial atoms and photons in cavities

Manipulating states of simple quantum systems has beome an important field in quantum optics and in mesoscopic physics, in the context of quantum information science. Various methods for state preparation, reconstruction and control have been recently demonstrated or proposed. Two-level systems/qubits and quantum harmonic oscillators play an important role in this physics. The qubits are information carriers and the oscillators act as memories or quantum bus linking the qubits together. Coupling qubits to oscillators is the domain of cavity QED and circuit quantum electrodynamics. In microwave CQED, the qubits are Rydberg atoms and the oscillator is a mode of a high Q cavity while in circuit QED, Josephson junctions act as artificial atoms playing the role of qubits and the oscillator is a mode of an LC radiofrequency resonator. The goal is to analyze various ways to synthesize non-classical states of qubits or quantum oscillators, to reconstruct these states and to protect them against decoherence. Experiments demonstrating these procedures will be described.

Description of a qubit

Any pure state of a qubit is parametrized by two polar angles theta,phi and is represented by a point on the Bloch sphere. A statistical mixture is represented by a density operator which can be expanded on Pauli matrices. The Bloch vector components are the expectation values of the Pauli operators. The qubit state is determined by performing averages on an ensemble of realizations. Overlap of two qubit states described by their Bloch vectors.

Manipulation and measurement of qubits

Qubit rotations are realized by applying resonant pulses whose frequency, phase and durations are controlled. In general, It is easy to measure the qubit in its energy basis. To measure an arbitrary component, one may start by performing a rotation which maps its Bloch vector along oz and then measure the energy.

Qubit rotation induced by microwave pulse

Coupling of atomic qubit with a classical field, microwave electric field linearly polarized with controlled phase. Qubit electric-dipole operator component along field direction is off-diagonal and real in the qubit basis.

α rotational

The rotating wave approximation/rωa neglects terms evolving at +-2wmw frequency. The rωa Hamiltonian is time independent.

Coherent state

Coupling of cavity mode with a small resonant classical antenna , the Hamiltonian for the quantum field mode fed by the classical source, Rotating wave approximation keeps only time independent terms.

Field Evolution

Field evolution in cavity starting from vacuum at t=0, we use Glauber formula to split the exponential of the sum in last expression. Expanding exp. In power series, we get the field in Fock state basis.

Estimation and reconstruction of quantum states in cavity: Fock and Schrödinger cat states

One might want to accumulates statistics about the

measurement of a complete set of observables, performed on a large number of realizations of the system. The results constrains the system density operator, which is found by solving a set of equations equating the observed statistics with the theoretical ones.

This procedure may lead to difficulties; if the data are noisy, it might happen that the direct constrained density operator is found to be non-physical, e.g. with negative eigenvalues. The number of available copies of the state might be small. The data presenting then large fluctuations. Often, the set of measured observables is incomplete, making it unlikely to constrain the parameters defining the state. In these situations, reconstruction of the state is an estimation problem:

How can we find the parameters defining the state from the information provided by incomplete measurements, performed on a finite set of copies and suffering limitations from noise?

Inspired by classical estimation theory, we will analyze the general method of maximum likelihood and maximum entropy before describing their application in cavity, illustrated by the reconstruction of Schrödinger cats and Fock states of a field.

Reminder about classical estimation theory
Fisher information & Cramer-Rao bound
The known probability law for the estimators depends upon an unknown parameter θ of one of the estimators which might also be a vector. For reasons made clear the probability function is called the likelihood of θ of one estimator corresponding to result of the other. The measurement of

probability brings an information about θ that we want to quantify. We are going to call estimator θx a function which associates to each x result an estimation of the true θ. The variance of θx averaged over measurements defines the estimator precision. This variance has a lower limit independent of the estimator called the cramer-Rao bound which is in turn related to a function of the likelihood called the Fisher Information. Measuring a random variable X yields a result x. The known probability law $p(x/θ$ of X depends upon an unknown parameter θ which can also be a vector. For reasons made clear the function $p(x/θ$ is called the likelihood of θ corresponding to result x. The measurement of X brings an information about θ that we want to quantify. We are going to call estimator θx a function which associates to each x result an estimation of the true θ. The variance of θx averaged over measurements defines the estimator precision. This variance has a lower limit independent of the estimator called the cramer-Rao bound which is in turn related to a function of the likelihood called the Fisher Information. First we consider unbiased estimators, whose average over a large number of measurements yields the true value of θ. We then use the identity $dp/dθ=p$ dLogp/dθ which leads to: $\int$ θ(x)-θt(p dLogp/dθt(dx=1

Then we square the integral and use the Cauchy-Schwartz inequality, thereafter comes the introduction the variance of θ and we get the Cramer-Rao inequality, where we have defined the Fisher information function of θ. Iθ would be the expectation value of the square of the logarithmic derivative with respect to θ of the likelihood function. The Fisher information definition is generalized to situations where θ is a multicomponent vector by introducing a Fisher matrix involving the expectation values of second order partial derivatives of $Log x, \theta$ with respect to the θ components., that is beyond the scope of this part.

Bayes law/Maximum Likelihood estimator

A natural choice for the estimator is maximised by bayes combined with some assumption of prior knowledge about the estimator θ before the approximation. The joint probability for finding values of couple estimators $p(x, \theta$ can be expressed in terms of the a priori probabilities over each of them, i.e. px and $p\theta$, If nothing is a priori known about one of them,let's say; θ, we should assume a flat $p\theta$ probability distribution leading to probabilities of each of them over their intgrated sum; $p(\theta/x = p(x/\theta) / \int p(x/\theta)d\theta$. The probability distribution of one of the estimators;e.g.,θ after result of the other; e.g., x has been found is thus given by the likelihood function $p(x/\theta$.

Photon count in cavity

To count up to nm photons we can either choose $\phi o = \pi/nm+1$ and use one detection phase ϕr corresponding for instance to the detection of σx and σy. After detecting p qubits in state j=0 and N-p in state j=1 the inferred photon number distribution has become the distribution maximum is obtained by computing the derivative of $p(n|p;N-p$ versus n. The derivatives cancels for X=p/N and the photon number nmax satisfies. To estimate the width of the inferred photon number distribution we compute its second derivative at X=p/N and we get the Taylor expansion of the photon number distribution around its maximum.

Simple illustration: a coin game

Consider a heads or tails draw, X taking the bit values x=0/1 with probabilities p and q=1-p, which we parametrize with an angle θ by defining $p=\cos^2 \theta/2$, $q=\sin^2 \theta/2$; we get $p(0/\theta=\cos^2 \theta/2$ hence

dlog $p(0/\theta)$ /d$\theta[^2=$tg$^2 \theta/2$;

$p(1/\theta=\sin^2 \theta/2$ hence dlogp$(1/\theta)$/d$\theta[^2=$cotg$^2 \theta/2$; and the Fisher information generated by a draw found to be θ-independent:

I$_1(\theta=p(0/\theta)$ [dlog p$(0/\theta)$ /d$\theta[^2+p(1/\theta)[$ dlogp$(1/\theta)$/d$\theta[^2=$ $\cos^2 \theta/2$ tg$^2 \theta/2+$ $\sin^2 \theta/2$ cotg$^2 \theta/2$

The precision of an optimal estimation of θ for N draws would be; IN=NI$_1$ thus $\Delta N(\theta=1/\sqrt{N}$ then deduce the standard deviation of p and q, and the standard deviation of X=$\Delta N(p=\Delta N(q=\cos \theta/2 \sin \theta/2$

Updating the field estimation: An exercise on Bayes logic

A field initial state in vacuum and being coupled to a resonant atoms entering C with no phase information, does not build any coherence between Fock states. Its density matrix thus remains diagonal in Fock state basis and the field quantum state is determined by its photon number distribution, you need to find out how p(n, is updated when a sample is detected.

The field updating is determined by the characteristics of the Ramsey interferometer. Let's recall the ideal conditional probability to detect an atom in j provided they are n photons:

πs ideal(j/n,=cos²(nφo-φr-jπ/2=1/2(1+cosnφo-φr-jπ

Due to imperfections, this ideal Ramsey signal is modified by offset and finite contrast and

becomes(b and c being calibrated in auxiliary experiments):

πs(j/n,=1/2(br+cr cosnφo-φr-jπ   br~1

Bayes tell you that if the atom is found in j, then the p(n, probability becomes:

P after(n/j,= πs(j/n, p(n,/Σπs(j/n, p(n,

P(n, is multiplied by the firnge function of the interferometer.

Comparison between Ramsey and Mach Zehnder interferometers

In Ramsey interferometer, two resonant pilses split and recombine atomic state in Hilbert; Qubit follows two pathes between R1 and R2 along which they undergo different phase-shifts. By final detecting of qubit in e or g and sweeping phase, one get fringes that informs about differential phase shifts between the states.

In Mach-Zehnder, the splitting and recombination occurs on particle trajectories. Beam splitters replace Ramsey pulses, Fringes inform about differential phase shifts induced on the pathes.

A special system: Circular Rydberg atoms coupled to a superconducting Fabry-Perot

Electron is localized on orbit by a microwave pulse preparing superposition of two adjacent Rydberg states: [e)→[e)+[g)

The local wave packet revolves around nuleus at the transition frequency between the two states like a clock's hand on a dial. The electric dipole is proportional to the qubit Bloch vector in the equatorial plane of the Bloch sphere.

The quantized JJ realizes a qubit

When the conjugate variables Q and ϕ are quantized, they become non-commuting operators. The product $Q\phi$ has the dimension of an action/energy*time, with the commutation preposition.

The linear LoC quantum oscillator has a ladder of equidistant levels separated by the energy hw. This equidistance is broken in the JJ system, due to the departure of the actual potential $\cos\delta$ from the δ^2 parabolic rule: The open JJ circuit is a non-linear oscillator. Due to the breaking of the transition degeneracy it might be possible to manipulate the two lowest states with microwave without exciting upper levels. Thus you can define a qubit whose frequency w_{01} close to w_{J1}, falls in the radio frequency domain/make an order of magnitude estimate of w with the values of e, I_0 and C.

In fact, this open circuit qubit is not practical because it is not controllable, you will see that by coupling it one can turn it into a manipulable device.

Magnetic effects: Flux quantizaion

The phase of the Cooper pairs wave function is constant in a small wire only in zero magnetic field. In a field B deriving from a vector potential A(r, the phase of the wave function acquires a gauge term taking the form: $\omega \propto \exp(-2ieA(r. r/h)$

The phase variation between two points become:

$\delta_1 - \delta_2 = 2e/h \int A.dl$

In a bulk metal, this condition combined with phase unicity requires that the curviline integral of A be proportional to the flux of B across the contour.

The SQUID: superconducting interferometer detecting weak magnetic fluxes

A superconducting loop wih two JJ a and b is crossed by a magnetic flux ϕ to be measured. The phase should recover the same value/modulo 2π after one turn. Calling δa and δb the phase jumps of the two JJ and assuming that the self-inductance of the circuit is negligible, you can get:

$\delta a - \delta b = -2\pi \, \phi/\phi_0$

when is fed by a constant current I, the circuit should fulfill the equation describing the interference between two Josephson currents.

$I = I_0 \sin(\delta_0 + \pi \, \phi/\phi_0(+ I_0 \sin(\delta - \pi \, \phi/\phi_0,$

By fixing I to $2\eta \, I_0$, the device operates in a superconducting manner if there is a value of δ_0 fulfilling the above equation. As soon as ϕ exceeds the threshold; $\phi_s = \phi_0/\pi$ Arc $\cos\eta$, the SQUID transits to the normal state and a voltage V develops between the two ports of the circuit.

Conditions to realize a superconducting qubit

We have described a JJ as a quantum system. Its variables p and δ, defined as macroscopic quantities pertaining to a system made of a large number of particles, are non-commuting operators, obeying to an evolution equation ruled by a quantum Hamiltonian. The phase defined modulo 2π is consistent with the discreteness of p which must assume integer values. In the systems described below (phase qubits), p will present large fluctuations and its discrete character will not be essential (this is different in charge qubits, not considered here). We will thus describe p and δ as continuous variables. The states of this quantum system have discrete energies and it is possible to isolate the transition between the ground state |o> and the first excited state |1>, which is non-degenerate with other tConransitions because of the JJ non-linearity. Restricting ourselves to exciting the o→1 transition with micowave pulses, you can force the system to evolve in the o-1 subspace, thus realizing a qubit. To get an operational system, we should include one (or several JJ's) in an electrical circuit in order to realize the following operations:

-Frequency tuning of the qubit
-Coupling of the qubit to microwaves in order to manipulate its state
-Coupling qubits with each other or with a microwave or radio frequency resonator to realize quantum gates
-Detection of the qubit with a state selective device

I now describe a device in that all these essential functions can be implemented:

The phase qubit cnt'd

When the JJ is fed by a dc current I produced by an external source, the current conservation in the circuit can be written by expressing the dc and ac Josephson prepositions:

$I = I_0 \sin\delta + dQ/dt = I_0 \sin\delta + C dV/dt = I_0 \sin\delta + hC/2e\, d^2\delta/dt^2$

This equation describes the acceleration of δ produced by the sum of two «forces»:

Proportional to $I_0 \sin\delta$, is the non-linear restoring force of the Josephson resonator, Proportional to I, is an applied force imposed by the source of current. The sum of these forces derives from a potential proportional to $-i_0\cos\delta - i\delta$. Comparing with the Hamiltonian of the open circuit JJ, we immediately get the current driven Hamiltonian: $H(i) = 2e^2/C\ p^2 - h/2e(i\delta + i_0\cos\delta)$; who defines the dynamics of a quantum effective particle with conjugate coordinates δ and p in a washboard potential.

The Josephson phase qubit

For I smaller than I_0, $U(\delta$, has minima around that it is quasi-harmonic and maxima. Let's focus on the system's dynamics around a minimum. The ground state and the first excited state in the potential well associated to this minimum are separated by frequency ω_{01}/a few GHz. The second excited state is linked to the first by a transition with frequency ω_{12} unequals ω_{01} due to the potential inharmonicity. So you can selectively excite the o→1 transition and realize an effective two-level system.

By increasing I, you can lower the barrier between two wells until you reach a configuration where state 1 has an energy just below the potential maximum. If the qubit is in state 1, the effective particle escapes by tunneling through the barrier and δ undergo an accelerated motion down the washboard. When $d\delta/dt$ exceeds a critical value, the junction transits to the normal phase and a voltage appears between its ports, which gives a detection signal selectively detecting the qubit in state 1. The state o remains stable in the well and undetected by this effect. In order to selectively detect o, you can transfer the system from o to 1 by a resonant microwave pulse and then detect state 1.

Tuning the phase qubit by varying current I

The frequency ω_{01} of the qubit depends upon the current I that controls the operating point δ_0 of the JJ and the shape of the washboard potential. The positions of the minima of $U(\delta,$ correspond to the cancellations of the force acting on the effective particle and are given by solving $I = I_0 \sin\delta_0$ or $\delta_0 = \mathrm{Arcsin}I/I_0$, with the additional condition that the second derivative of $U(\delta,$ is positive. At these points, the system is in equilibrium and the relation $I = I_0 \sin\delta$ means that the current go across the intrinsic JJ without component across the capacity. In the vicinity of this equilibrium, we have:

$U(\delta, = U(\delta_0, + (\delta - \delta_0)^2/2 \; d^2U\delta_0/d\delta^2 = Cte + I_0 \, h\cos\delta_0/4e \; (\delta - \delta_0)^2$

By manipulating the qubit with a variable current I one can thus tune its frequency and detect its state by lowering the height of the barrier. These operations are carried out in sequential terms. I now describe a variant in which the control and the detection are performed by an inductive coupling of the qubit to external circuits.

Controlling the phase qubit by the flux

Instead of controlling the qubit by a dc current, one can do it inductively by sending a magnetic flux ϕe across the circuit. In practice, ϕe is produced through a superconducting dc transformer coupling the qubit circuit to an external one. M is their mutual inductance and L is the classical self inductance of the qubit circuit. The controlling current $I\phi$ produces the flux $\phi e = MI\phi$ across the circuit qubit. An induced current I appears in L that opposes the incident flux, producing a total flux:

$\phi = \phi e - LI$

Applying the phase quantization relation, we get the condition satisfied by the JJ phase:

$\delta = 2pi \ \phi e - LI/\phi o \ ; \ \phi o = h/2e$

that yields the expression of the magnetic energy of the L inductance versus δ:

$\frac{1}{2} LI^2 = \phi^2 o/2L \ [\delta/2pi - \phi e/\phi o[^2$; and leads after adding the capacitive and intrinsic inductive contributions, to the Hamiltonian of the qubit controlled by the flux ϕe:

$H = 2e^2/C \ p^2 + \phi^2 o/2L[\delta/2pi - \phi e/\phi o[^2 - \phi o/2pi \ Io \ cos\delta$

By changing ϕe you vary the phase of the cosine at the minimum, which changes the shape of $U(\delta,$ When $2pi \ Io \ L/\phi o$ becomes larger than one, $U(\delta,$ has a double well shape. The potential is symmetrical for

$\phi e/\phi o=1/2$, dis-symmetrical otherwise. The symmetrical solution corresponds to the flux qubit.

For a convenient choice of L, the well knows a small number of bound states, the two lowest ones forming the qubit. Detection is made by tunnel effect, as for the current controlled qubit, by adjusting the flux to vary the height of the barrier.

The system leaks toward the deeper well that has a big density of excited states in the vicinity of the barrier summit. By finely adjusting ϕe you can also tune the qubit frequency/change of the potential curvature.

Order of magnitude of phase variance

By developing the qubit Hamiltonian up to second order in δ-δo, you can obtain the harmonic oscillator expression:

$H \sim E_c\ p^2+E'j/2\ (\delta-\delta o)^2$; whose angular frequency close to $\omega o1$, is:

$\omega oh \sim \omega o1=1/h\ \sqrt{2E_cE'J}$

The variance of δ in the ground state is given by:

$E'J/2\ (\delta-\delta o)^2=h\omega o1/4$

By elimination of E'J between the last two equations also considering that $E_c=2e^2/C$, you then get a simple expression linking $v o1=\omega o1/2pi$; the flux quantum, the junction capacity and the variance of δ:

$(\delta-\delta o)^2=e/\phi oCv o1$

A typical qubit frequency $v o1=5GHz$, you can have $\Delta\delta=\sqrt{(\delta-\delta o)^2} \sim .13$ radi. The phase is relatively well defined, hence the name phase qubit while the conjugate variable variance is large: $\Delta p \sim 1/2\Delta\delta \sim 5$.

Detecting the phase qubit with a SQUID

When the qubit transits from one well to the other, δ changes by about pi which corresponds to a change of about Io of the current in the qubit circuit and to a flux variation of about LIo~ϕo.

This sudden flux jump of one flux quantum is detected by a SQUID inductively coupled to the qubit (a voltage appears between the ports of the SQUID when the qubit transits). The flux controlling circuit (flux bias) is used to finely tune the qubit frequency and bring it suddenly to the threshold of selective detection of its quantum states at the time of measurement.

Manipulating the qubit state with rf pulses

An ac current -Irf sin(ωt-ϕ, with frequency close to ωo1 can be used to excite the qubit across a coupling capacitor and to realize rotations of its Bloch vector. The qubit Hamiltonian becomes:

H(t)=HQB+ϕo/2pi δIrf sin(ωt-ϕ,

The interaction term, proportional to the position δ of the effective particle representing the qubit is reminiscent of the electric dipole coupling term in an atom. You can develop the sine function keeping only the resonant term/RWA approximation and move into the frame rotating at angular frequency.

Capacitive coupling of two qubits

Consider two identical phase qubits A and B. You can couple them by a capacitor whose capacity Cx is very small compared to the capacity C; that of the JJ. The voltages VA and VB on the two sides of Cx are 2epa/C and 2epb/C respectively/pa and pb are the canonic momenta of the two circuits. The coupling energy induced by Cx:

Hint=1/2 Cx[VA-VB[^{2}=2e^{2}Cx/C2 [PA-PB[^{2}

By grouping this term with the two qubit hamiltonians, the total would be:

HA+HB+Hint=H'A+H'B+H'int; where

H'i=[2e^2(C+Cx)/C^2[p^{2}i+ U(δi,   et   H'int=-4e^{2}Cx/C^2 pApB

The qubits mass is slightly renormalized which modifies their common frequency and a coupling term appears which is proportional to the product of the canonical momenta. This coupling lifts the qubit degeneracy and produces a frequency doublet in the spectrum of the coupled systems.

The open circuit JJ Hamiltonian

Expressing the two Josephson relations as canonical equations deriving from a Hamiltonian H, we get:

dp/dt=-Io/2e sinδ=-dH/hdδ; dδ/dt=4e^{2}p/hC=dH/hdp

H=2e^{2}p^2/C- h/2e Iocosδ

H defines the dynamics of a non-linear oscillator whose momentum and position are p and δ. The combined dc and ac Josephson effects induce an oscillation:

204

a phase difference δ induces a current/dc effect. This current produces a charge imbalance that creates, by capacitve effect, a potential across the JJ. This voltage induces in turn, via the ac Josephson effect, a variation of δ. This produces a coupled oscillation of the phase and the charge. For small δ such that $\cos\delta \sim 1 - \delta^2/2$, the oscillator behaves linearly, its linearized Hamiltonian H1 being:

$H_1 = 2e^2p^2/C + h/4e\, I_0\delta^2$

The quantum feedback

Quantum feedback can be implemented in Cavity QED to prepare and stabilize Fock states against quantum jumps. Two methods have been tried: The first, the actuator is a classical source injecting small pulses of coherent radiation in the cavity. The corrections of n = ±1 quantum jumps are achieved by incremental steps made of pulses with positive or negative amplitudes, many pulses of decreasing intensity being required to make the field converge back into a Fock state.

The transient off-diagonal density matrix elements generated in the process are destroyed by the quantum collapses induced by the dispersive probe atoms. The process takes a few tens of milliseconds, making the procedure relatively slow and impossible to implement for n>4. In the second method, the actuators are single resonant atoms able to inject or subtract a photon in one step, making the procedure more reactive and faster. Fock states up to n=7 have been prepared and protected in this way. Extending the method to protect other kinds of states, such as Schrödinger cat states is an interesting field of investigation.

Applying quantum feedback to the stabilization of Fock states
Fock states are interesting examples of non-classical states;
They are fragile and lose their non-classicality in time scaling
as $1/n$. The preparation by projective measurement is random.
The crucial part here is to consider:
1 Would it be possible to prepare them in a deterministic way
by using a quantum feedback procedure?
2 Can these procedures protect them against quantum jumps
(loss or gain of photons)?

Feedback procedure
-Inject an initial coherent field in C
-Send atoms one by one in Ramsey interferometer
-Detect each atom, projecting field density operator in new
state estimated by computer
-Compute displacement which minimises the distance between
target and new state
-Close feedback loop by injecting a coherent field with
amplitude alpha in C
-Repeat loop

Fock state reconstruction: Max Ent vs Max Like
We prepare an $|n>$ state by running a sequence of POVMs
realized with probe qubits atoms. We then displace it by $-\alpha i$
(different values distributed on lines passing by phase space
origin). Then we measure sigma x or sigma y on ~10 atoms

with a phase shift $\phi_0 \sim \pi/2$ per photon. We average over ~100 to 200 realization for each α_i. The same data are used to obtain ρ by Max Ent and by Max Like. For Max Ent we average the difference of count rates in j=1 and j=0 for each α_i value and we look for the optimal ρ reproducing these averages under the exponential form. For Max Like, we use the first 3 atoms out of the about 10 crossing C after the field displacement and we measure for each α_i the frequencies $f_i(p)$ for detecting 3-p atoms in j=1 and p in j=0. This we neither exploit atom count averages, nor atom correlations on a single realization.

Single atom prepares Schrödinger cat state
1. Coherent field is prepared in C
2. Single atom is prepared in R1 in a superposition of state e and g
3. Atom shifts the field phase in two opposite directions as it pass through C: Superposition leads to entanglement in typical Schrödinger cat situation
4. Atomic states mixed again in R2 maintains cat's ambiguity:

$|\uparrow,e> + |\downarrow,g> \ (|\uparrow> + |\downarrow>) \ |e> + (|\uparrow> - |\downarrow>) \ |g>$

Detecting atom in e or g projects field into+or- cat state superposition.

Steps in preparation and reconstruction a cat Initial state: $|g>$
R1:$\pi/2$ phase shift Injecting a coherent field ß Displacing field by $-\alpha$ R2:$\pi/2$ phase shift
Phase-shift $d\phi(n)/dn$ (in units of π) versus photon number

207

Taking into account light-shifts non-linearity, We choose a small Delta value to get a large phase-shift per photon. The non-linear terms in n in the expansion of the atom-field states make phase-shift per photon n-dependent: about π for n = 0, it is ~ 0.5π for n=5.

Effects of the non-linear phase shift
1. The cat prepared by the 1st atom would be distorted
2. It Modifies the direct reconstruction procedure

Reconstructing Schrödinger cat states by Max Ent
Since the measured observables are «close» to parity, they are «almost binary» and the Max Ent method applies well. We have perform the NG= 500 field displacements and measure the expectation values of the corresponding errors affected generalized parity operators (with one phase ϕr). Since the measurements do not change n, we use in each realization the information provided by~10 atoms which reduces the number of realizations necessary per displacement. The searched ρME is the exponential of a linear combination of 500 Gi(errors) operators. The coefficients of this combination are 500 Lagrange multipliers. These multipliers are determined by a least square fit minimizing the X^2 sum given in the discussion of the Max Ent method . The theoretical curves superimposed to the experimental peak points are fits obtained with the amounts of these multipliers. The agreement

between the experiments and the fits is quite good. Once ρME has been determined, we compute the true W. We thus go from W(gen)(errors), given by the direct data, to the true W by indirect route.

The Schrödinger cat decoherence

The cavity mode gets entangled with the reservoir. This entanglement is responsible for decoherence. As soon as the two environment states get orthogonal, there is a «which path» information in the environment that lifts the quantum ambiguity of the state superposition and destroys the quantum coherence. The field density operator at time t is obtained by tracing over the environment, as long as the two field components remain orthogonal to some extent.

The review of measurement theory

A Standard/von Neumann Measurement

Determined by giving the ensemble of projectors on a basis of eigenstates of an observable G represented by a hermitian operator. Finding the probability of the result i on a system in the state ρ before measurement and the state ρi in which the system is projected by the measurement, the number of projectors is equal to the Hilbert space dimension/2 for a qubit, and the measurement is repeatable: After a first measurement, one finds again same result, as a direct consequence of the orthogonality of the projectors. Besides these standard measurements, one might want to define POVM measurements.

POVM

A POVM is defined by an ensemble of positive hermitian operators Ei having non-negative expectation values in all states realizing a partition of unity. The number of Ei operators can be arbitrary, either smaller or larger than the Hilbert space dimension. The POVM is defined by the rules giving the probabilities standard measurement rules. Since the Ei are not normalized projectors, one might find the POVM procedure a statistical measurement since it yields a result belonging to a set of values, with a probability distribution.

POVM realized as a standard measurement on an auxiliary system

A POVM on a system in a Hilbert space can always be reduced to a projective measurement in an auxiliary system belonging to another space B, to which the system is entangled by a unitary transformation. Let us associate to each element i of the POVM a vector |b> of B, the |b>'s forming an orthonormal basis, i.e. Space B has a dimension at least equal to the number of POVM elemnets, and let us consider a unitary operation acting in the following way on a state |ω>A |o>B, tensor product of an arbitrary state of A with a reference state |o>B of B. This operation conserves scalar products and is thus a restriction of a unitary transformation in A+B.

Max Like reconstruction principle

Inspired by classical estimation theory, we would like to infer the most likely density operator ρ reproducing the statistics of N observations of a quantum system given in N identical copies. We list all the POVM's Ei elements associated to the measurement results, each POVM element being labelled by its index; We express the result of a complete measurement by the list of all the i's, each being associated to its frequenc of occurrence fi=ni/N. If I is a continuous parameter, we discretize it in bins. The theoretical probability pi of finding the result I when the state is ρ writes:

Pi=Tr[ρEi,

85Rb nF7/2 Rydberg states using purely optical detection

This work demonstrates the first frequency measurements of rubidium Rydberg levels using a purely optical detection scheme. The Rydberg states are excited in a heated Rb vapour compartment and Doppler free signals are detected via purely optical means. All of the frequency measurements are made using a wavemeter which is calibrated against a GPS disciplined self-referenced optical frequency comb. We find that the measured levels have a very high frequency stability, and are esp. robust to electric fields. The apparatus has allowed measurements of the states to an accuracy of 8.05MHz. The new measurements are analysed by extracting the modified Rydberg-Ritz series parameters.

The accurate measurement of highly excited Rydberg level energies in the alkali atoms plays an important role in improving the accuracy of atomic models. In most Rydberg spectroscopy experiments the atoms are detected via field ionization. However, in this study we use a method of purely optical detection in an ordinary vapour compartment. A vapour compartment is a convenient and straightforward solution for finding Rydberg levels, that could potentially permit rapid advances in Rydberg spectroscopy. This technique presents a method of finding Rydberg states quickly, with a large signal to noise ratio and an apparent insensitivity to electric fields, which makes it particularly suited to studying high ℓ Rydberg states with large polarisabilities. It is therefore important to verify the ability to perform precision spectroscopy in such a setup. Though there is a large body of work on precision interval and fine structure measurements of the different rubidium Rydberg series, measurements of the energies of these levels are more difficult to carry out, and are therefore mainly limited to the lower ℓ states. It appears that measurements of the 85Rb nF series have only been made once by Johansson in 1961 for n=4-8. However, as new tools are now available in laser spectroscopy, such as the optical frequency-comb technique, it is interesting to return to such measurements. Precision laser spectroscopy measurements of Rydberg states could be effectively made using purely optical detection with a vapour compartment sample. During the experiment nF7/2 Rydberg states between n=33-100 were excited in 85Rb using a three step laser excitation scheme. The three step level system consists of a 780.24nm transition 5S1/2 F=3 to 5P3/2 F=4 , a 775.98nm transition 5P3/2 F=4 to 5D5/2 F=5 and finally a 1260nm transition 5D5/2 to nF7/2. To observe excitations to Rydberg states, the first two step lasers are fixed at their respective transition frequencies and the absorption of the 780nm laser is monitored whilst the 1260nm laser is swept across the transition. This technique involves the quantum amplification effect; due to the large differences in decay lifetimes of the three excited states of the system, the excitation of a single atom by the third step laser will hinder many absorption-emission cycles on the second step transition. This in turn will hinder a large amount of cycles on the strong first step cycling transition which can cause a measurable decrease in the first step absorption. Optical pumping is applied on all three steps with σ+ polarised light. Optical pumping on the first step transition ensures the second step laser only excites to the $mF=5$ sublevel of the $5D5/2$ $F=5$ hyperfine state. Therefore the third step laser can only excite a single transition, the $5D5/2$ $F=5$ to $nF7/2$ $F=6$. Having a well defined pathway to the Rydberg states is important because of the relatively small 10MHz splitting of the $5D5/2$ level.

In the experimental setup used for measuring Rydberg state frequencies, the first step is phase locked to a self-referenced optical frequency comb and the second step is frequency locked using a separate rubidium reference compartment. The first and third step laser light is transported to the comb and wave-meter using single mode optical fibres. The first two steps are circularly polarised using quarter wave plates, and the third step laser is circularly polarised using a broadband Fresnel rhomb. All three lasers are focused to a beam waist of 100m inside the compartment, which increases the available third step laser power density. The vapour compartment is heated to a temperature of 60 C to ∘ increase the atomic density in the compartment and to therefore enhance the first step absorption.

Before adding the third step laser to the system, we verified that efficient optical pumping was occurring on the first step transition by scanning the second step laser across the $5D5/2$ manifold, with the first step laser locked. The first step laser selects only zero velocity atoms, and therefore the second step laser scan showed a single and symmetric Doppler free peak in the first step absorption. This single peak, with a FWHM of 11.5MHz, corresponds to the reduced absorption of the first step laser as the second step laser excites the $5P3/2$ F=4 to $5D5/2$ F=5 transition. By adding a small frequency modulation to the second step laser, and monitoring the first step absorption via a lock-in amplifier, an error signal is extracted. Using our frequency comb we verified that this second step frequency lock was repeatable to an accuracy of 1MHz on a daily basis. We found that it is possible to detect lower n states with a very good signal to noise ratio. Therefore to verify the line shape of the detected third step transitions the photo-diode was monitored directly on an oscilloscope during a fast scan a cross the $5D5/2$ to $33F7/2$ transition. The scan was carried out and the frequency axis was calibrated using Fabry Perot resonator at 1268nm. To improve the detection sensitivity of third step transitions, a frequency modulation is added to the third step laser via the injection current, with a modulation amplitude of 15MHz and frequency of 6kHz. Detection of the first step absorption is carried out at the first harmonic using a lock-in amplifier with a time constant of 1 second. The free running third step laser is scanned by applying a linear voltage ramp to the laser Piezo using computer software and a Digital to Analogue converter interface. The free running laser stability was measured as less than 1MHz over one second, which is sufficient to carry out slow scans across the Rydberg transitions. We found that the wave-meter stability stayed below 2MHz for times of 1000s. We also ~ found that the wave-meter was able to maintain accuracy of 6.2MHz across the 1254nm1268nm range, when regularly calibrated at 780nm. Therefore, throughout this experiment the wavemeter is calibrated every 30 minutes to the comb-locked first step laser, to supply a direct frequency link with the comb.

The Results

The third step transition absolute frequencies were collected for n=33-50 in intervals of one, and from n=50-100 in larger intervals of five. Fitting to the transition data was done using a Wahlquist first derivative function. We found that the line widths of the detected third step transitions prevented resolving the $nF7/2$ and $nF5/2$ fine structure splitting in this experiment, which for n=33 to 100 is 4.35MHz to 0.16MHz respectively. However, the use of σ+ light for the third step laser ensures only the $nF7/2$ level is excited in this case. Ten traces were taken for each state in order to understand the repeatability of the measurements. It was found that on average the standard deviation of each set of ten scans was 2MHz with an accuracy limited by the short term drift of the

215

wave-meter during the time taken to collect each set.

To study potential frequency offsets of the transitions caused by power shifts, pressure shifts or Zeeman shifts we took measurements of both high and low n states with a range of different first, second and third step laser powers, compartment temperatures and opposite circular polarisations respectively. We also checked for errors from time delays in the data acquisition process by scanning the third step laser across the same transition in opposing directions. No repeatable shifts of the transition frequencies were found with increased laser powers or compartment temperature and therefore potential offsets from these effects were not added as corrections but instead the spread of measurements were used to estimate a maximum error in each individual case. Neither Zeeman shifts nor time delay errors were detectable within the short term accuracy of the wave-meter and therefore these effects were assumed to give a negligible contribution to the uncertainty. To measure potential DC Stark shifts of the Rydberg states we applied electric fields of up to $30Vcm^{-1}$ across the vapour compartment and checked for frequency shifts of both the $33F7/2$ and $100F7/2$ transitions. In each case there was no measurable deviation. This unexpected observation was also made in references when detecting Rydberg states in a compartment. A screening of the Rydberg atoms inside the compartment seems to be present, which makes them resilient to electric fields. This is a very positive effect as it allows precision spectroscopy of high ℓ states with no DC shifts.

Max Like reconstruction principle

It is useful to introduce the log of L(ρ , C being the combinatory factor:

Log L(ρ, = C+ NΣfi Log Tr(ρEi

It satisfies the general inequality

Log L(ρ, smaller than C+NΣfi Log fi, that stem from the relation, true for any set of positive xi/concavity of Logarithm

Log Σxifi larger than Σfi Logxi   (fi larger than zero, Σf1=1,

Setting xi=Tr(ρEi(/fi in last equation and taking into account the first equation and Log Σi Tr(ρEi(=1

leads to the inequality for Log L(ρ, and sets its upper-bound. Determining the maximum likelihood solution amounts to finding the ρ operator such that Log L(ρ, gets as close as possible to the satisfying the relations: Tr(ρEi= fi, that correspond to the best possible coincidence between the measured frequencies and the theoretical probabilities.

Qubit Tomography by Max Like

Let's estimate by Max Like the density operator of a qubit defined by its Bloch vector:

ρ=1/2(1+ΣPj δj,   j=x,y,z

The six POVM elements Ej+- with the frequencies fj+- that we have found lead us to the Logarithmic of the likelihood function.

Haroche

References

[1, M. Mastriani, *"Quantum Boolean image denoising"*, Quantum Information Processing, Springer US, pp.1-27, 2014. (DOI) 10.1007/s11128-014-0881-0

[2, Mastriani, M.: Quantum Edge Detection for Image Segmentation in Optical Environments (2014). arXiv:1409.2918 [cs.CV]

[3, Weinberg, S.: The Oxford History of the Twentieth Century (Michael Howard & William Roger Louis, editors ed.). Oxford University Press. p. 26, (1998)

[4, Weinberg, S.: Einstein's Mistakes in Physics Today. Physics Today 58(11), 31, (2005) doi: 10.1063/1.2155755

[5, Koashi, M., Imoto, N.: arXiv:quant-ph/0101144

[6, Bohm D.: Quantum Theory. Prentice-Hall Inc. 1951.

[7, The Nature Of the Universe, A series of Broadcast Lectures by Fred Hoyle, Oxford 1952

[8, Zitiert nach: Kragh, Helge: Cosmology and Controversy. Princeton University Press, New Jersey,1996

[9, Gamow, G.: Expanding Universe and the Origin of Elements, in: Physical Review, S.572-573, 1946

www.ingramcontent.com/pod-product-compliance
Lightning Source LLC
Chambersburg PA
CBHW070853180526
45168CB00005B/1802